Seo

A No Nonsense Guide in Digital Marketing

(Learn How to Drive Traffic and Boost Your Business in the Digital Age)

Patrick Bohner

Published By **Zoe Lawson**

Patrick Bohner

*Seo: A No Nonsense Guide in Digital Marketing
(Learn How to Drive Traffic and Boost Your
Business in the Digital Age)*

ISBN 978-1-998927-87-6

No part of this guidebook shall be reproduced in any form without permission in writing from the publisher except in the case of brief quotations embodied in critical articles or reviews.

Legal & Disclaimer

The information contained in this book is not designed to replace or take the place of any form of medicine or professional medical advice. The information in this book has been provided for educational & entertainment purposes only.

The information contained in this book has been compiled from sources deemed reliable, and it is accurate to the best of the Author's knowledge; however, the Author cannot guarantee its accuracy and validity and cannot be held liable for any errors or omissions. Changes are periodically made to this book. You must consult your doctor or get professional medical advice before using any of the suggested remedies, techniques, or information in this book.

Table Of Contents

Chapter 1: Local Seo And Small Businesses

So, what is close by seek engine marketing and advertising, and how can you benefit from it as a small community enterprise? Google relentlessly works at making seo the most effective, precious, and rewarding technique for every person who wants to be seen on line. You can be a multinational business enterprise, a social commercial enterprise enterprise, a media house, a business company entity, or an individual— you want to be visible and heard.

Everybody can't use the identical approach or equipment to advantage visibility on line. The seek engine advertising technique implemented thru a small-scale close by industrial organisation as a way to develop is probably one among a type from seo for an global corporation. Local search engine optimization is essentially a narrowed or a niche version of search engine optimization.

As the selection suggests, close by search engine advertising includes the geographic element of a are seeking, specifically meant for organizations in close proximity to a capability customer. Besides the location of the consumer, network search engine marketing specializes in the relevance and prominence of the purchaser question on the internet.

For instance, if a person searches for "vegan consuming locations close to me," the Google seek engine will display them a list of vegan ingesting places nearby. Even despite the fact that there is probably plenty of particular vegan ingesting places within the path of the town, Google will display high-quality the ones which might be without troubles handy based totally mostly on proximity to the searcher.

Prior to having seo capabilities, human beings could have been searching for high-quality locations, recollections, merchandise, or services thru visiting miles to shop for them truely because they could not find what they have been seeking out of their very very own community. Local search engine optimization permits nearby organizations to find their customers easily. It is good for each sellers and consumers, because it saves a extremely good deal of cash, fuel, and time.

With the strength of community seo, purchasers can get very specific with their are looking for phrases. For instance, you can search for "fantastic bars near me" or "pleasant stay activities this night," and Google will show you the ideal are searching for effects. So, as a nearby commercial enterprise corporation, you need to discover your cue from what customers typically kind in the search bar and insert the identical keywords and key terms on your content fabric cloth to seem excessive in search scores.

How Should You Apply Local seek engine advertising in Your Business Strategy?

To start your adventure with close by seo, you want to have positive topics in vicinity. You need to make the extraordinary use of content material advertising and marketing, clever technology, and capabilities to be had. Google has made matters fairly generation-driven for each agencies and customers. It's almost like Google reads your thoughts, is acquainted at the side of your conduct, and brings to you things which you intend to search for.

Local are seeking for engine marketing and advertising and marketing has acquired recognition with the heightened use of cellular telephones. Now human beings have the choice of without difficulty attempting to find a few element they need on their telephones. Therefore, as a small close by industrial business organisation, you should take advantage of the modern-day character-pleasant era and Google's clever set of rules.

Here's a checklist at the manner to follow on your small community corporation to discover its marketplace:

Local Citations

First, ensure you listing your commercial enterprise employer in on-line nearby directories, collectively with Google My Business, Facebook Pages, Yelp, Yellow Pages, Instagram for Business, and Apple Maps, in fact to call a few.

You need to encompass your neighborhood commercial organization call, deal with, and speak to amount in order that your customers can contact you effects. Also, make sure that your touch information is the equal in all of the community directories and it's miles current.

As you construct your profile, you want to additionally write some lines about your organization highlighting its essential dreams. Google will pick out up the ones keywords and deliver customers your manner. You have to moreover encompass the hyperlink for your internet internet web page if you have one. Not all small companies have their net sites prepared when they first open, that's super. You can however be a part of the ones online directories.

Local citations are imperative for your rating on Google. So, ensure which you pick great genuine directories like Google, Facebook, and plenty of others., to earn credibility.

Your Google Business Profile

You should have a organisation profile on Google. That's your stepping stone inside the direction of embarking on a journey of a small network organization entity. It permits you to function your company statistics to Google Search and Google Maps.

On your business agency profile page, you embody your bodily cope with, cellular telephone range, website link, workdays, operational hours, and a brief description of your business agency. You can also be asked to pick out out your enterprise class absolutely so your corporation can be listed consequently. You may moreover get the opportunity to add some pictures of your enterprise's products and services, this is really a exceptional manner to intensify your business organisation's selling factors.

Customer Ratings and Reviews

In nowadays's digital technology, client opinions play a large role in building or ruining your organization reputation. Therefore, you want to take this difficulty of your community are searching for engine marketing critically.

Respond to reviews right away, whether or now not or not they're proper or lousy. If you need to improve your ratings, ensure you reply to awful reviews and take them in stride. Don't ignore them. You should express regret in reaction to a poor assessment; try and understand your patron's mind-set and locate strategies to enhance your services.

As prolonged as you interest on giving your customers the extremely good product, carrier, or revel in and cling to their powerful remarks, you'll have higher scores and evaluations through the years.

Create Hyperlocal Content on Your Website

You can lure your customers through a hyperlocal content material approach, it is to embody area-precise key terms on your content fabric. The cause of hyperlocal advertising and marketing is to increase the foot web page site visitors of a quite centered nearby purchaser base. It's an effective technique, as maximum folks that make hyperlocal searches intend to shop for a specific services or products. Therefore, it's vital for a network business employer owner to maintain their net internet site on line optimized with hyperlocal seek engine marketing.

Besides placing the right key terms for your net website content material cloth cloth, you should issue out your vicinity to your URLs, add area-specific metadata (you may look at metadata similarly within the e-book), optimize for voice are trying to find, and upload your contact information.

The Benefits of Local seek engine advertising for a Small Business

A network search engine marketing marketing and advertising and marketing strategy has many rewards for small businesses furnished they apprehend clients' options and look at the techniques Google uses to list the brilliant available retail options for nearby searchers.

Take some time to have a look at and exercise all of the seek engine marketing satisfactory practices that permits you to make your commercial enterprise business enterprise internet web page internet

internet site on-line rank first on the are searching for engine.

Below are a number of the reasons why studying and utilising community are seeking for engine marketing is a need to for small organizations.

A Cost-Effective Marketing Strategy

Local search engine advertising allows you gain your customers without a good deal monetary investment. There isn't always any want with a purpose to rush into marketing campaigns, create brochures, or positioned up billboards for your organization to be seen to humans. You can sell your neighborhood company pretty successfully with Google. Not to say, it furthermore saves you a wonderful deal of transport and packaging expenses.

Improves Your Visibility Online

Local are seeking engine advertising brings your business enterprise records to the top of the hunt effects, which allows you advantage on-line visibility. The brilliant element is that the visibility you get is valuable because it's frequently your functionality customers who see your commercial enterprise statistics in the are looking for effects. The more you appear in the pinnacle seek consequences, the higher visibility you can get through Google.

Reach Customers Who Are Ready to Buy Your Product or Service

Since nearby seek engine advertising is based totally on searcher's proximity and reason, it's distinctly probably for a community commercial organization to get a sale. Google offers choice to local searches. There are lots of folks who make "near me" searches, which reason them to unique neighborhood organizations around them. Even on the identical time as humans do not kind "near me," Google no matter the reality that suggests them your enterprise within the are looking for consequences. Why? It's because of a function called geotagging (identifying your location), which you need to have on your website on your clients to find out you.

Stay Ahead of Your Competitors

With more herbal leads and customer traffic coming through on-line searches, you can live beforehand of your opposition. They may additionally additionally have the same products or services as your business, but in all likelihood they may be the usage of 1/2-baked marketing and search engine optimization strategies.

Google changes its set of guidelines steady with purchaser behavior. More and more humans use their cell telephones to make cause-based totally completely searches and area-unique queries, which lets in internet web sites with hyperlocal search engine optimization display up on Google Maps. Such nearby corporations also can appear in Google's three-Pack list, which shows the pinnacle three community outcomes. A pinnacle community search engine advertising pastime plan builds your authority and recognition on-line similarly to offline.

Chapter 2: Why Do Some People Fail At Search Engine Optimization?

SEO is a huge exercising. It might not be all people's cup of tea. While some human beings are able to take their companies to fantastic heights with seo, there are numerous who fail at it. Search engine optimization is a way that is developed constantly primarily based mostly on what people look for on line.

Succeeding or failing with search engine marketing is based upon on whether or not you're capable of understand the purpose in the lower back of a query, key-phrase, or key word. It's all approximately observing and studying why human beings use serps like google within the first location.

seek engine marketing may be overwhelming for some people. It calls which will have persistence and examine it each day. You can't in reality advantage a immoderate organic score on Google in an afternoon or . Many people don't recognize that it calls for consistency and staying energy to gain authority with Google or with any searching for engine for that keep in mind.

It takes time to bring together top content material material this is helpful, applicable, informative, and interesting. Besides, there are such an entire lot of various factors that decide your score on search engines like google and yahoo.

So if you have surely kickstarted your search engine optimization slog as a small community corporation and you don't need to fail at any charge, in reality hold in mind the following elements.

Lack of Knowledge

As you start dabbling in are searching for engine advertising, take into account to advantage a few facts first. It's no longer realistic to expose off your commercial enterprise company on-line without strong information of what works and what doesn't paintings on-line. You can start sluggish, however you want to assemble your information every day.

Read plenty about pleasant search engine optimization practices to develop your understanding at the undertaking. Pay heed to what exclusive groups on your vicinity of hobby are doing in terms of promotions and campaigns, the type of content fabric they put up, and their method toward their clients.

Not Staying Updated on Algorithm Changes

Don't be ignorant approximately serps like google' set of policies adjustments. Instead, stay abreast of era updates and the way Google aligns itself to such improvements. Keep your eyes and ears open to what Google is up to.

Perhaps the maximum vital detail to realize about search engine advertising and marketing is that it goals at giving searchers the awesome feasible stop result for their question. It is important that you understand how people search for data. Is it on their cellular phones in particular, is it thru voice are searching for for, or are they searching out movies? The set of rules is based on purchaser rationale and conduct.

Ignoring the Important Aspects of search engine optimization

There are numerous factors to search engine optimization, and it's essential to combine all of them into your advertising and advertising roadmap. Search engine marketing is not pretty an awful lot keywords. That's only a small a part of it. You begin with the proper keywords and then skip without delay to different key elements of search engine optimization—name tags, meta description, photograph optimization, inner linking, outside linking, internet website loading pace, cell responsiveness, and the patron-friendliness of the website.

Poor Quality Content

Content is the spine of any search engine optimization blueprint. Your content cloth has to help humans in a few manner. It need to be relevant to a patron question. If your content cloth material doesn't fulfill a searcher's want, it's of no want. Good content material material constantly gives accurate and complete statistics. There are companies that don't have any content cloth to inspire human beings to buy their services or products.

Unnecessary Keywords

Keywords are essential, however they want to now not be overused. Your attention have to no longer be on setting key phrases in a placed up. Do your critical key-phrase research earlier than growing content material. However, allow the preferred key phrases mixture inside the content material seamlessly. Google doesn't admire too much key-word stuffing.

Not Being Consistent

Consistency is prime to success. We all realise that, right? Your seo efforts need to be regular so as on the way to see a few outcomes. Google rewards web net sites with everyday exquisite content material. You should make modifications, corrections, enhancements, and updates in your corporation internet web page or internet website frequently, and character interaction wants to be happening.

Being inconsistent gives your opposition an opportunity to supersede you.

Lack of a Strategic Approach

You can't acquire your industrial enterprise enterprise goals without a strategic technique. It's quite clean. When you're visible on-line, you are competing with innumerable comparable businesses or entities. So what need to you do to face apart?

You need to have a strategic way to advantage your search engine optimization desires. For search engine advertising and marketing to be just right for you, there desires to be some thought and method within the returned of it. You want to investigate your preferred overall performance often, provide you with new mind to enhance visibility, preserve your self updated with the contemporary day updates, and practice search engine optimization techniques inside the right manner.

Not Keeping Track of Analytics

You want to recognize what's occurring backstage, such as ways plenty organic traffic your website gets each day, what number of conversions (real shopping for) are taking region, what key phrases fetch the maximum internet page visitors, and what the deliver of your net internet site on-line site traffic is. The better you recognize your internet site's analytics, the higher motion you may take toward enhancing your net site's performance.

Not Giving It Enough Time

Many human beings don't supply seek engine advertising and marketing and advertising enough time to paintings for them. It's now not constantly about horrible SEO. Most of the time, humans surely do matters 1/2-baked with out a exquisite deal records or know-how and assume consequences. Well, it's now not possible. You have to deliver yourself definitely to a task to appearance it blossom. Your searching for engine marketing game plan is like a challenge—a method to help you gather sure goals in your agency. So offer it time.

Chapter 3: Essential On-Page And Off-Page Seo Strategies

As a small nearby commercial agency, you have to apprehend seo in its totality and create your advertising and advertising method for that reason, every on-internet net page seo and stale-page search engine advertising. In clean terms, on-page are searching for engine advertising and marketing is everything you do on your net internet web site, and off-net page seek engine advertising is the movement you are taking outdoor your net internet site online to sell it.

For example, you write enticing content fabric with hyperlocal key phrases to pull in more clients to your product. Then you furthermore can also add some quality optimized pix to go together with it. That's your on-net page seo. However, seo does now not give up there.

You additionally need to make sure which you are making sufficient off-internet page searching for engine marketing efforts to unfold the phrase about your industrial company. How does that display up? That happens even as you positioned up about your products on social media or gain a few links from others in your net website, which helps you rank better on Google.

So allow's study every on-web page and off-web page seek engine advertising for your neighborhood enterprise to develop.

On-Page seo

On-web web page search engine advertising and marketing and advertising is a direct way of gaining visibility based totally on are attempting to find results. It can be associated with the kind of content material material fabric you create, the key phrases you use, the facts about your industrial corporation, or maybe the technical are trying to find engine advertising and advertising involved inside the method. It's basically your personal natural efforts to succeed with Google.

Good Keyword Research

The first step closer to developing amazing content material cloth material is seeking out applicable key phrases. As a small industrial agency, you have to be looking for community key terms geared toward customers internal close proximity. For instance, if you are a grocery keep proprietor and you want human beings within the equal vicinity to understand about your save, you want to select out key phrases like "nearby grocery shop," "nearest grocery keep," or "closest grocery hold."

Quality Content

You want to create beneficial content material cloth to your internet website that consists of all of the vital details about a selected topic. It ought to solution all character queries, resolve individual troubles, and guide them to make a attempting to find choice.

Title Tags, Meta Descriptions, and Image Optimization

After you have written appealing content material fabric with relevant key phrases, now could be the time to make sure that it's optimized and prepared for Google to move slowly your web page. You need to encompass the primary key terms that you would really like to rank for in your understand, URL, and your meta description.

The meta description is a -line precis of your entire post, which appears on seek engine effects pages (SERPs). Since it's the meta description that a searcher sees first, you want to make it extremely good attractive for them to click on on on through the hyperlink and go to your net internet site on line. You write a meta description for each submit manually in the meta description field.

Images play a massive position in fetching you some correct site visitors and moreover influencing people to shop for your products or avail of your services. However, it's essential which you optimize all of your photographs earlier than which include them. Name the picture and compress its length. You additionally have the possibility to allow are trying to find engine crawlers to expose your net pages with photographs, so that you have to embody the primary key terms on your photograph call and decrease the scale for better tempo.

User-Friendly Website

Your internet website online ought to be clean to navigate and need to no longer appear cluttered. Your content material should flow systematically for readers to apprehend information with out trouble. For a community industrial agency net site, it wants to be mobile-best. Most people look for groups the usage of their phones, whichmakes it vital that you optimize your internet internet page for cell indicates.

Add Internal and External Links

Many human beings don't apprehend the fee of which includes inner and outside links to their content. You should hyperlink to other applicable pages of your very personal net site. For instance, if you have written a publish on "why vegan meals is healthful," you can point out some vegan merchandise that you promote and link to the 'products' web page.

Also, add at least one first-rate outside hyperlink to your positioned up, which is relevant on your put up. For example, when you have written a weblog about "the way to preserve a wholesome way of life," you may hyperlink to a brilliant informative article on "applicable consuming behavior."

High authority outdoor hyperlinks increase your internet internet web site's reputation.

Off-Page search engine optimization

Off-net web web page search engine advertising is an oblique manner of gaining visibility on are looking for outcomes. It can be via link building, social media advertising, or brand mentions. Although off-net web page are trying to find engine advertising and marketing techniques moreover play a fantastic function to your Google ranking, it's now not lots for your manipulate.

Link Building

The first element you need to popularity on in terms of off-web page seek engine advertising and marketing and advertising is constructing your link profile. The easy approach to attain this is with the useful resource of developing content material material cloth that people would possibly need to link to. Of route, it's no longer so clean. However, it could be finished over a time period.

Your content fabric must be terrific powerful and first rate for brilliant web sites to pass on their hyperlink on your net internet page, which means their vote to your content cloth.

Link building is critical because engines like google like google supply priority to net pages that are encouraged via pretty some people. It's like amassing votes to your content material material. The extra humans are on your choice, the much less complicated it is going to be to be able to acquire the pinnacle.

It's continuously greater beneficial to accumulate hyperlinks from unique assets. Securing hyperlinks from one large website, over and over, is not going to assist hundreds. The better deal is probably to accrue links in a exclusive style.

Social Media Marketing

You have a remarkable possibility to sell your commercial enterprise on numerous social media channels, that could mean higher visibility, greater website visitors, greater customer engagement, and greater income. It's a win-win!

Brand Mentions

As you begin developing your on line presence, growing content material material fabric, and selling your merchandise on specific systems, you may get discovered via numerous human beings. You earn logo mentions via severa people. Your customers ought to mention you of their social media posts, or influencers and bloggers can also want to point out you, which leads to more sales. That's how your off-net web page seo works!

Customer Reviews

Perhaps one of the most great techniques to assemble your on-line popularity is via securing real purchaser critiques. You want to continuously inspire your clients to go away opinions on Google Maps or for your product page. Customer critiques convey some of weight. They can earn you lots of latest clients and in reality a higher rating on Google.

Word-of-Mouth Referrals

When humans endorse your services or products to others, it's referred to as word-of-mouth referrals. This may be considered off-internet page are trying to find engine advertising advertising and advertising and advertising and marketing. So, in no way underestimate the electricity of old school, word-of-mouth selling of your logo. In reality, you want to invite your clients to refer your services and products to their friends and friends.

Guest Blogging

You can write visitor blogs for others and gain some of their site visitors on your internet net website. How does that artwork? You propose to provide appropriate great content cloth to a excessive authority internet page (web sites that already rank immoderate with Google), this is just like your area of interest, and ask them to link once more on your website. It will now not simply provide you with greater web page site visitors, however it'll moreover help you enhance your net net web page authority.

Both on-web web page and rancid-net internet page search engine advertising and marketing are critical for your success as a small industrial organization. You can take gain of the hyperlocal advertising approaches and become the authority to your locality.

Chapter 4: Organic Marketing Or Paid Advertising

Should you choose paid marketing or stay with natural techniques of advertising? It's a question that you want to ask your self, then do an assessment on what's tremendous on your business company. There are severa alternatives to get greater eyes on your services or products absolutely through spending a few extra money.

You can take benefit of paid advertising and advertising and marketing. You can set your personal charge range, spending as masses or as low as you want. It relies upon in your advertising and marketing rate variety and the way a long manner you want to move in regard to promoting your commercial corporation.

It is in truth essential to offer your emblem a similarly push in nowadays's competitive situation. Many people are promoting the same topics that you are attempting to rank for. It's particularly vital that allows you to be visible anywhere so that you can acquire leads and generate more earnings.

However, it's no longer so smooth. Paid advertising and marketing has its drawbacks. It want to now not be overdone. You ought to no longer be excessively promotional or appear spammy. Building your brand organically is appeared as if it'd be too tough. Placing paid advertisements on various structures is taken into consideration greater effective.

Yes, paid advertising and marketing or marketing is a notable preference. As lengthy as you do it efficiently, it can have amazing results for you. There are quite some advantages of setting paid commercials on Google and social media:

●It can be pretty price-powerful. In an preference like pay in step with click on on (PPC), you pay simplest when your commercial is clicked on.

•It is a quite focused manner of advertising and marketing. You get to area your commercial in the front of your potential customers.

•It is a brief way of conducting your customers, and you've were given whole manipulate over your commercial.

Paid branding does have its blessings. However, the advantages come thru best when you have a way in area. You cannot actually location advertisements randomly with out a great deal notion within the back of them. It's an art work to benefit leads and conversions via advertising.

You moreover want to undergo in mind that classified ads are disturbing to maximum human beings. When you region an ad, it doesn't look like a regular put up. Words like 'Sponsored' or 'Ads' are stated on top, which also can discourage human beings from clicking them. Most of the time, such posts are ignored or marked as spam.

On the other, an herbal way of selling your brand is constantly more first rate and useful ultimately. People are able to construct don't forget in manufacturers that offer their merchandise in a herbal way without being glaringly promotional.

Organically advertising your products or services can be a slow way, and it may get a touch worrying at times whilst you don't see income going on. Nevertheless, the right form of herbal advertising done with clever search engine advertising and marketing techniques ought to have prolonged-lasting rewards.

There are many benefits of selling your brand with a way-driven plan:

●You advantage an actual target market. People who like your sponsored submit or click on on on on a paid advert won't be genuinely inquisitive about your product. They ought to probably have given you a like or a click on while casually scrolling a internet site. On the other hand, even as a person likes your normal publish, they're much more likely to be interested in your product.

●You don't spend any money in any respect, and it brings you achievement.

●It permits you construct brand recognition. You have the electricity to create interest approximately your products or services thru developing exciting posts on social media.

●It accentuates your function as an tremendous logo.

●The amazing element is that it brings regular and everlasting site visitors in your net website.

You need to determine whether or not you want to pick out paid advertising and marketing in your small commercial enterprise or in case you want to paintings strategically to your content material and your interaction with your customers. You can meet your commercial agency goals thru making use of every the techniques— herbal similarly to paid. However, you want to apprehend and study the experts and cons of each.

Chapter 5: Local Seo Tools For Your Small Business

The global of digital advertising and marketing is charming. As you still boom your small business employer with Google, you may discover loads of factors to enhance your popularity as a close-by entrepreneur. All you need is the attention to research and put into effect generation in your gain.

There are severa neighborhood seek engine advertising tool available that will help you music the increase of your nearby commercial enterprise enterprise, apprehend your competition, find out search engine optimization-pushed key terms, diploma your general performance, and provide you with new schemes and thoughts as a manner to hold thriving as a small commercial employer on your neighborhood network.

So which seek engine advertising device should you operate?

Since there are so a number of them, it can get a piece overwhelming to select out the proper ones in your close by seo efforts. Here's a list of the extraordinary community search engine optimization equipment in your small organisation.

Google My Business

As we have said in advance, to begin your nearby search engine optimization pastime plan, you need to installation your commercial enterprise business enterprise profile on a Google My Business account with all your organisation statistics. Google My Business offers you the electricity to locate your community competition, examine organizations, and acquire purchaser critiques.

Besides, it's far quite a customer-exceptional device. Your clients will see a characteristic known as "e-book an appointment," in an effort to nudge them to find out your listings. You also can change or replace your information as and when required. It's an easy way to keep your clients updated approximately your products or services and furthermore your street deal with, enterprise organisation hours, cellular phone range, and internet site URL.

Google Maps

Small businesses can extract a whole lot of benefits from the Google Maps app. It can act as a awesome digital marketing tool. If used cleverly, Google Maps can do a lot extra than increasing your visibility in your viable clients. It also may be used to region your enterprise the proper manner in your ability clients. Now, how does that art work?

Google offers your business organization precedence on seek results if you have given your specific region with none errors or missing information. You need to furthermore make sure that your deal with is proven to your Google Maps listing. Another thing to keep in thoughts is which you have to check all instructions of your commercial organization listing, as a manner to help Google recognize you better and rank you better. You want to be as unique as viable along side your commercial employer facts.

Remember, community seo is a totally vicinity of hobby search engine advertising and marketing and advertising and marketing. For example, if you run a sports sports garb save, point out unique key phrases like "sports sports clothing keep." You need to now not include indistinct key terms like "strolling footwear" or "song pants for running."

Use suitable exceptional, properly-optimized pics for Google Maps. People need to get a experience of the kind of location you've got got, and that could inspire them to go to it.

Google Search Console

It's important to track your net net page common overall performance occasionally. You need to be privy to any technical issues that might be hampering your website's ranking. Google Search Console is a exquisite tool that will help you abide through the high-quality necessities of Google. You can use it to track your net pages that Google has indexed, the key phrases or queries that your internet site ranks for, the clicks and impressions, and if it's technically appropriate for cell customers or now not.

Google Ads Keyword Planner

The approach of attempting to find the proper key terms on your close by agency can be tough. However, Google Ads Keyword Planner can assist you decide on the relevant key terms in your enterprise corporation to advantage the pinnacle spot on are looking for outcomes.

You should be able to find the hunt terms which may be normally typed in via manner of searchers of a big range. The remarkable factor is that Google Keyword Planner is a loose device that shows you the possible key phrases in your situation matter further to the quest extent for every one in every of them. It's a powerful region to start to your boom activity plan and pay in line with click on (PPC) campaigns.

Moz Local

Moz Local is a relatively effective tool from a nearby seek engine advertising mindset. It saves a number of your time and effort with the useful resource of assisting you create and manage your neighborhood commercial organization listings on Google, Facebook, and unique channels under one dashboard.

You don't need to fear approximately any duplicate listings because of the fact Moz Local will delete them automatically. The tool optimizes your profile for better community visibility and attain. It gauges photos and information required for your profiles and listings.

Moz Local additionally allows on your popularity control via preserving track of client reviews all through exquisite structures. It indicators you as quick as there can be a new client compare simply so your reaction time is brief. Negative feedback is handled properly to preserve customers.

Whitespark

Whitespark gives you extra Google-centric effects. It helps your community agency rank higher on Google Search via monitoring your ranking and focusing on nearby searches. You can assemble your nearby citations without difficulty. The tool will discover all of the locations your commercial enterprise profile is indexed, your competition' citations, and it's going to test and screen your citations' boom.

BrightLocal

BrightLocal is a whole device that allows you to execute all of your key advertising and advertising and marketing efforts from one place. You can recovery your community seek engine advertising and marketing troubles and make better improvement with its community are seeking audit function. It also has competencies like Google My Business audit, popularity control, quotation tracker, and network seek rank checker.

GrowthBar

GrowthBar is a fantastic device for taking your search engine advertising sport a stage up. The device permits you to diploma outcomes out of your content material fabric cloth advertising and advertising and are searching for engine advertising, find out all those relevant key terms which you concept were tough to find, and hold a watch fixed steady to your competition. It additionally comes with a free GrowthBar Google Chrome Extension, which offers you get admission to to seo insights sooner or later of your Google searches.

* * *

These device and software program beneficial resource you in augmenting your seo technique. However, you need to pick out out a device that suits your enterprise dreams and allows you construct your emblem within the pleasant feasible way.

Chapter 6: How To Use Social Media To Build Your Brand

Doing seek engine marketing right is most effective one facet of the coin. The different facet is to construct your brand focus, and this is going beyond your searching for engine advertising and marketing and advertising strategies. Good close by search engine optimization advertising ought to be the inspiration of your emblem. However, to take it ahead, you need to increase a completely particular voice to your emblem.

There may be a dozen similar agencies on your locality, and they is probably making use of the equal seo techniques as you. What is it that you want to do to live in advance?

The secret's to build your logo interest.

There are numerous mediums to spread the word approximately your products and services. Of course, you need to do it in a significantly strategic way; otherwise, it obtained't be powerful.

Promote Your Business on Social Media

You want to harness the energy of social media advertising and marketing. If you do it successfully, it can generate splendid effects for you. First things first, you need to recognize your product—its marketability, relevance, and price proposition. The higher you recognize your product, the less complicated it gets to promote it.

The idea isn't to be indistinct or random approximately your advertising and advertising technique. Social media can located your products or services inside the front of such masses of customers, which means that more leads and more conversions.

Why Use Social Media for Your Small Business?

Although social media doesn't have an impact on your Google are attempting to find engine ranking at once, it helps within the following strategies:

●It gives greater publicity to your internet web site. You can add your internet net site hyperlinks on your social media posts, and it will force some internet web site traffic your way.

●It boosts your brand consciousness, brand recognition, and popularity.

●You get a risk to connect to a latest purpose marketplace.

When you percentage some facts about your product or services in an interesting manner on social media, your fans are probably to reshare it on their feed, so that you can growth your visibility. The greater your logo is suggested via the usage of others on social media, the more credibility you will earn over a time period.

Social media posts improve your net net site search engine marketing and advertising in an oblique way. When a person clicks through your internet web site link on social media, it's an intentional click. The individual is aware about what they will be going to locate as they land in your net internet site online. They are pretty probable to spend some time exploring your products and services on your internet website online. When someone spends exceptional time on your internet site, visits numerous pages, in all likelihood exciting them to do so, it influences your seek engine advertising and marketing in a outstanding manner.

So what want to your social media marketing method be for your small enterprise?

Choose the Right Social Media Channel

To begin with, you need to choose the right channel to your brand merchandising. It cannot be in reality any social media platform, as every social media channel has its non-public one-of-a-kind purpose. For instance, Facebook is meant for personal, more emotion-driven content material material. It's a place in that you stay in contact collectively with your own family and friends. Twitter is for records and evaluations. Instagram is for innovative people, and LinkedIn is for professional connections. So you need to choose out out accurately the proper region to sell your corporation.

Facebook, Instagram, and YouTube are fantastic systems for logo constructing. Facebook is remarkable for small nearby businesses as such a variety of human beings use it.

Be Disciplined About Your Posting Schedule

Your fulfillment on social media is depending on how constantly you positioned up. Therefore, you want to have a social media calendar so that you agenda your posts earlier and stay consistent. It's continuously higher to create a bank of content material so that you in no manner compromise in your posts' brilliant.

You need to have your duplicate, pictures, movement pics, hashtags, and hyperlinks planned earlier.

Stay Engaged and Interactive

You want to not simply positioned up content cloth on social media—you need to additionally have interaction with others. Create content fabric cloth that instigates humans to interact. Ask questions and create polls and quizzes. Reply to feedback on your posts, write feedback on others' posts, and like and percentage top content fabric through others to energy more traction for your internet web page. Social media is all about being social.

Focus on Building a Community

As a small neighborhood enterprise organization, you need to attention extra on gaining a devoted network of enthusiastic lovers in place of a large variety of ignorant fanatics. Create content fabric that could encourage, assist, or add fee in people's lives. Remember that people don't like promotional posts, so chorus from selling your brand an excessive amount of.

For instance, in case you sell vegan food products, don't leave your product hyperlinks in all of your posts. Instead, create posts across the advantages of vegan food or instigate conversations around a vegan way of existence. You will mechanically construct a vegan network over a time period.

Get Into Collaborations, Giveaways, and Events

Social media is a powerful medium for on line collaborations, giveaways, and occasions. Now the query is, who've to you collaborate with and the way? You need to be part of palms with other organizations that aren't your direct competitors. For instance, if you run a eating place, you can collaborate with a person who has a shop promoting special meals merchandise. You can run campaigns on Facebook and Instagram, tagging each special to offer exposure to every of you.

You can plan giveaways and sports together. Think of espresso or meals-associated contests wherein your espresso-loving followers would like to participate. Offer to offer free espresso samples to the winners. Cross-promoting within the form of contests is a awesome manner to stir up enthusiasm and hobby among the ones who've a have a look at you. Not to mention, you'll gain new fans thru such institutions.

Have a Unique Brand Voice and Stick to It

Let your business enterprise business enterprise have a one-of-a-type voice, and permit that be obtrusive in all you do on social media. To find out your emblem voice, you need to apprehend your services or products and your target market. For example, in case you run a fashion garb maintain and your audience is in maximum cases the younger crowd, you need to gather your logo voice that resonates with the younger era.

What is it which you do in a amazing way? Why are clothes at your keep greater appealing? Think of your particular promoting factors (USPs) and permit them to come upon. Post images of your branded packaging every so often for humans to connect with it. Share photos and motion photographs that spotlight the sort of garments you sell.

Chapter 7: How To Stay Ahead Of Your Competitors

In the previous financial disaster, we determined out a way to use social media as a tool to build your emblem as a small neighborhood enterprise employer. In this financial disaster, we're able to learn how to stay ahead of the competition. When you're on foot a business enterprise, you could have opposition spherical which you want to cope with. It's now not sensible to brush aside the present opposition. On the opposite, you ought to stay updated on what's taking area round you.

Your competition help you enhance as a agency, as they keep you in your feet. In reality, your opposition push you to avoid errors of any type. They push you inside the course of excellence.

Implement the Best Local search engine optimization Practices

Doing everything effectively in phrases of appropriate nearby search engine optimization—content-based absolutely are seeking engine advertising and marketing, era-primarily based seo, and social media search engine optimization—is crucial on your enterprise in in recent times's virtual age. It's furthermore one of the pinnacle approaches to overcome the competition in the market.

For example, you are the proprietor of a health club, in which you offer personal jogging shoes, health coaches, and weight-reduction plan plans for your customers. With the proper are searching for engine advertising implementation and by means of adapting to enhancements in generation, you may surpass your competition. Your fitness center is without issues searchable on Google Maps, your industrial business enterprise info are updated at the Google My Business web page, you've got got were given a kickass content advertising and marketing and marketing approach in place, and also you in reality have a tremendously engaged social media following.

Know Your Customers

You ought to realize who your customers are—operating professionals, homemakers, innovative individuals, artists, technicians, or commercial enterprise human beings. To supersede your opposition, you need to apprehend approximately your clients and apprehend who your target market is.

In nowadays's speedy-paced age, companies want to have an opening approach in the direction of their advertising techniques. Your commercial company can't be about the whole thing, and it could't be for all people. It wants to have a targeted patron base. You need to cognizance in your customers—their wishes, alternatives, likes, and dislikes.

Know Your Competitors

It's essential to recognise your opposition, too. In reality, it must be a part of your business business enterprise technique. You need to discover what your competition are as much as in terms of new product launches, how they address their customers, and the way crucial they're with their are trying to find engine marketing strategies.

If your competition are performing some problem particular, you need to take a cue from them and expect of latest and modern mind for boosting your products or services.

Provide Excellent Customer Service

The high-quality way to live in advance of your competition is to delight your clients. You need to be ideal in conjunction with your customer support. Even one example of awful customer service can motive a awful assessment, that can tarnish your brand picture.

Angry customers and awful opinions can change your boom graph. Therefore, the proper aspect to do is to maintain a terrific relationship at the aspect of your customers. Always add a personal touch in your issuer. For instance, in case you run a eating place, attend to your clients' dreams in my opinion, make sure your body of workers is prompt and aware of their requests, and ask them in the event that they cherished the meal.

A suitable organization is all about satisfied clients, which ends up in extra customers and more income.

Train Your Employees

A actual emblem usually has its values and guidelines instilled in its employees. Provide your workforce with important schooling once in a while for them to execute the right approach in the course of your enterprise desires. With the right form of training, they will recognize their roles higher and make a contribution in your earnings.

Analyze Your Sales Data

You ought to have a look at your income records periodically and look at your non-public simple performance. If it doesn't appearance on pinnacle of factors, think about strategies to enhance profits. Maybe you want to artwork harder to your virtual strategies or marketplace your services or products in a greater powerful way.

There are severa sides to marketing and advertising and reaching out for your clients for higher visibility and profits. Find out your very own weaknesses and overcome them.

Let Mobile Technology Help You Accelerate Growth

Mobile technology plays a key function in accelerating growth for small organizations. Almost absolutely everyone nowadays has get proper of access to to smartphones, which makes subjects clean and brief for every groups and customers. As a nearby entrepreneur, you can do masses with the generation to be had on your smartphone and go away all opposition at the back of, such as:

•optimizing content material fabric and images on Google Maps

•with out problems supplying your business enterprise organisation location to clients

•enhancing employee average overall performance due to coping with client requests through mobile phone, saving effort and time

•ensuring smooth and smooth communique among sellers and customers

Stick to Your Brand Values and Ideas

The awesome tactic to hold your commercial organization in advance of your competition is to stay right in your logo values. Your customer need to recognize you for who you're. Make fantastic you keep consistency on your content fabric advertising sports activities in your website and social media. Your content material material cloth must be real and significant. It need to supply out your emblem philosophies simply so your reliable clients revel in associated with you and suggest your products and services to others as nicely.

As an amazing logo, you should constantly attempt to supply a few aspect extra in your customers. It can be inside the form of beneficial content fabric, precious information and steering on a topic, or an unexpected act of kindness. For instance, if you very very very own a salon, you can percent tips on physical grooming on Facebook. Let your social media posts display how your salon seems after customers on a non-public stage and gives solutions each time desired.

Besides, you have to be bendy and typically be open to recommendations, enhancements, and changes.

Chapter 8: The Black Hat Are Trying To Find Engine Marketing Tactics To Avoid

The recreation of search engine marketing is not truely restricted to doing the right seo, but it's additionally approximately no longer doing the wrong subjects. There's something referred to as black hat are searching for engine advertising and marketing, which means that that the wrong are seeking for engine advertising and marketing practices. As a nearby agency, you need to be privy to these practices and absolutely keep away from them.

The easy rule of correct are seeking engine advertising and marketing is to create outstanding content material material. To make your content material material handy to its site visitors, you want to study amazing search engine optimization tips defined with the beneficial useful resource of Google. However, if you do seo in a manipulative way, it's now not desired via way of Google in any respect or through way of any are in search of for engine for that be counted. In reality, you will be penalized for black hat seo practices, that could advocate your internet pages may be removed from Google Search, and you'll lose all herbal website online traffic. Nobody may additionally need to need that to take vicinity, proper?

So, so that it will maintain your difficult artwork and see its end result, you need to live a ways from any form of black hat search engine advertising and marketing techniques. Keep in thoughts that Google goals at giving its searchers the pleasant result viable. There isn't any way you may cheat or control the set of guidelines.

What are the ones black hat processes?

Keyword Stuffing

Some marketers anticipate that in the occasion that they add more keywords to their content fabric, it's going to assist them climb the ladder of ranking quicker. But it's up to now from the fact. You are intended to use key terms in a sure way and in high pleasant places for your internet site. For example, when you have written a publish titled "10 Ways To Stay Healthy," the primary key phrases want to be within the pick out, URL, and meta description, preferably placed somewhere on the start of the submit. However, the rest of the key phrases ought to be spread thru the placed up in a herbal manner.

When you use key terms manifestly, it allows clients find your put up, and that they gain from it. On the alternative, even as key phrases are honestly stuffed without a good deal cost, the visitors don't find out the content material fabric beneficial and will depart your internet site.

Cloaking

Cloaking may also moreover furthermore sound a chunk technical to you, however you want to understand approximately it. When someone provides particular content material cloth or URLs to search engines like google like google like google and human customers, it's termed cloaking. For instance, this will be giving a web page of HTML text to search engines like google like google and yahoo whilst displaying a web page of pics to searchers. It's a contravention of Google's Webmaster Guidelines, and you need to keep away from it the least bit charges.

Duplicate or Automatically Generated Content

This can be quite common. Duplicate content cloth fabric is honestly copying content material cloth from somewhere and reposting it as your very very own for your net internet page. Automatically generated content fabric approach copying content material material fabric with outstanding key terms in it and rehashing it without including lots fee to it. Thin content, low-first rate content material fabric, and spammy content material fabric are all black hat search engine advertising practices to benefit score with out putting in the real work.

Paid Links

It's pretty a not unusual exercising among many content fabric creators and digital companies to ask for paid links. They accomplish that to govern their page rank. Paying cash or providing a free product to a person to get a "do-have a look at" hyperlink from them isn't an superb exercising. Also, too many link exchanges, leaving unsolicited mail remarks on popular blogs, or stepping into traveller positioned up advertising and marketing are all the wrong hints to acquire fulfillment on search engines like google and yahoo like google.

Fake Redirects

In certain situations, redirects are required. For example, you've got were given moved your internet site to a modern address, otherwise you've consolidated particular pages into one. However, redirects which can be misleading to every search engines like google and customers want to be prevented. For instance, while a are attempting to find engine crawler receives to look a incredible web page on the identical time as a searcher gets a very special page displayed, it is a fake redirect.

If you need to do a legitimate redirect, do it cautiously with 301 redirects or JavaScript redirects. You might also moreover need to get assist from a technical professional to do this.

Doorway Pages

Doorway (additionally known as gateway) pages are created to rank rather for remarkable seek terms. Such pages aren't beneficial to customers, as they will grow to be stumbling upon similar pages with now not something valuable to them.

When human beings create pages virtually to rank for keywords in choice to to help customers or remedy their problems, it's a black hat method that should not be practiced.

The fantastic manner to score a high ranking on Google and on line reputability is thru ordinary first rate content material fabric and using most effective white hat search engine marketing and advertising methods.

Chapter 9: How To Use Google Analytics As A Marketing Strategy

Should you use Google Analytics on your community business corporation's advertising and marketing and advertising method? Well, yes. If you pick out out to stay unaware of your net net site analytics, you're lacking out on an entire lot of power in regard to the usage of the boom of your small commercial enterprise organization. Google Analytics also can furthermore appearance complex inside the beginning. However, it's tremendously useful and directional for your content material cloth approach and business enterprise desires.

Google Analytics serves the following key functions for a small commercial enterprise. It:

●measures your internet website on line internet site visitors

●indicates your website on-line web website online traffic' area

●indicates resources of your net net page website online traffic

●comprehends customer conduct

●lists the maximum visited pages of your internet net page

●highlights the most well-known content material on your net website

●gives you statistics of the visitors who convert into leads or customers

●offers insights into advertising and advertising and marketing thoughts that deliver greater site visitors for your website

●tells you in case your net web site is technically apt for cellular cellphone customers

If you would like your small employer to be determined through way of manner of its target marketplace, analyzing Google Analytics of your website is the modus operandi. It permits you keep away from effective advertising and marketing and advertising and marketing errors, that you are probably to make with out knowing your net website's analytics.

How to Use Google Analytics

It's pretty clean to put in Google Analytics and installation your account. If you've got already got a Google account, you may sincerely go to Google Analytics and sign in. You may be asked to feed in facts about your net website, on the side of your account call, net website online name, URL, and company class. After you have got were given finished filling within the records, you may be given a Google Analytics monitoring code that you will be seeking to installation on every web page of your net internet site on line. The tool of putting in the tracking code may additionally moreover need to differ for one-of-a-type structures, topics, and plugins. If you locate it overwhelming to do it yourself, you should get assist from a technical professional.

What's Next? Set Up Your Goals

Installing Google Analytics makes experience fantastic if you have positive business corporation desires in vicinity, and you tune your boom once in a while to enhance your search engine optimization approach.

Google Analytics lets in you to set up desires, which includes internet website online website visitors, individual engagement, sort of people signing up, and conversions. The most important detail to preserve in mind is how a bargain your internet site is helping your commercial agency to make bigger. Therefore, you need to be tracking the following metrics periodically, month-to-month or quarterly:

●conversion funnels—what number of clients are going thru the ranges of making a purchase

• conversions by using using their assets— wherein your searching for customers are coming from

• range of latest patron critiques

• geo-centered key-word clicks, that allows you to come up with an idea of man or woman purpose and interest in seeking out your products

• facts for Google My Business, which include clicks, impressions, requests for instructions, and call calls (extra consumer queries will bring your business company web page up in Google's Local three-Pack)

• man or woman visit duration in your net website—the longer someone remains in your net web website online, the better risk they will be real clients

As you start to have a look at and examine Google Analytics, it's crucial which you recognize what you need to music and degree in regard on your small close by business enterprise. Obsessing an excessive amount of approximately internet website net web page perspectives isn't always clever. You might also additionally reap masses of common visitors from all forms of belongings. However, you want to prioritize its relevance.

You need to cognizance on:

- place-unique visits

- patron cause

- your goal marketplace

You want to preserve song of business business enterprise-associated movements, collectively with queries, shape fill-ins, signal-ups, and so forth.

A clever entrepreneur will study topics holistically. However, they may apprehend what to reputation on and what to do to enhance organisation-driven metrics.

If you see your key metrics aren't appearing nicely, you want to get lower returned to running on your nearby SEO, social media marketing, product positioning, and technology.

Google Analytics will display you the path to correction, improvement, and boom. It's up to you the way you adapt to it and set the proper KPIs on your small business.

Chapter 10: How To Master Google Ads And Online Advertising

Promoting your small enterprise with Google Ads and online marketing and advertising is one of the treasured factors of your everyday advertising avenue map. You want to be aware about its blessings. Even despite the fact that natural marketing is superior to paid advertising and advertising and marketing and marketing (communicate to Chapter four), Google Ads are relevant and significant for your corporation.

You want to be strategic about how you use Google Ads. Consider its attributes, features, and advantages in advance than you make a decision to put money into on line advertising.

How Google Ads Work

Google Ads showcases your services and products on top for viewers to look and make a purchasing for desire. There are particular forms of Google Ads:

•Search Ads are the classified ads you take a look at on Google seek results.

•Display Ads seem on considered one of a kind places of a net web page.

•Video Ads are visible on YouTube or Google Display Networks (diverse websites, apps, and films in which Google Ads can be placed).

•Shopping Ads, due to the fact the call indicates, are meant for stores.

Based for your business's advertising and advertising and marketing desires, you can choose out appropriate ad campaigns.

Benefits of Online Advertising With Google Ads

Target-Driven Campaigns

With Google Ads, you may create extremely region of interest campaigns that target your goal marketplace. Such commercials have keywords to purpose clients who want your services and products. They are based totally at the consumer's region, age, language, and kinds of gadgets. The classified ads are verified at high-quality hours or days, and you may furthermore decide the frequency in their show.

Quicker Results

Google Ads gives quicker outcomes than are trying to find engine advertising and marketing. Of route, natural SEO can create prolonged-time period fulfillment for anyone. However, in case you want to acquire your customers speedy, on line marketing accomplished in a logical manner can yield superb outcomes.

Higher Reach

You can aim at extra key phrases, as a way to will allow you to obtain extra of your capacity clients who are not your lovers. The extra humans click on on for your ads, the better visibility and attain your corporation receives.

Boosts Brand Awareness

Your logo may also even get extra publicity, as your commercials can be displayed in the the front of recent audiences who is probably interested by your services and products. They may even get a hazard to find out your logo and its robust factor, that's specially probably to show into extra emblem mentions and hints.

Measures Success Easily

With Google Ads, individual interaction is in fact apparent. You understand whilst a person clicks for your ad, makes a buy, or browses via your products. Since you are aware about every valuable movement through manner of the use of customers, it is simple to degree achievement. You can prevent campaigns that don't be just right for you and repeat the ones who've completed properly.

Manages Your Campaigns

You can manage all your campaigns with beneficial gear like My Client Centre (MCC) manager account with the resource of Google. It allows you to keep tune of all of your campaigns in one location, which saves time. You can also change or edit your campaigns multiple instances.

Controls Your Budget

The fantastic element approximately Google Ads is which you have entire manage over your expenses. You pay handiest even as a consumer clicks for your advert. You can decide how an lousy lot you want to spend steady with month, consistent with day, and consistent with ad. There isn't always any minimal value, so it gives you entire freedom and flexibility.

Now on the identical time as you understand the blessings of the usage of Google Ads, you want to moreover take a look at the technique to apply it effectively.

How To Master Google Ads

Mastering Google Ads is all about using the right strategies to make your campaigns a achievement. The best approach to this entire ball game of online advertising is to position your product efficiently.

You want to hold the following elements in mind for your product positioning or brand positioning:

●It may be a great-based or price-based absolutely logo positioning. Your ad campaigns can spotlight that your logo sells incredible merchandise at less high priced prices.

●It can be a hassle-solving positioning. You can role your commercials as a manner to your customers' problems. For example, you want to promote it your plumbing services in a locality in which there are hundreds of plumbing troubles.

●It might be a advantage positioning, wherein your campaigns can highlight the benefits of your products.

●It is probably cost-based totally positioning, wherein you keep effective requirements for your products. You sell them at immoderate expenses because they may be remarkable merchandise, and you may make the customers accept as real with so.

Online advertising and marketing has won recognition through the years. More and extra corporations have paid advertising as part of their promotional techniques, which is why it's pretty competitive and tough.

The nice manner to rise above the market competition and no longer fizzle away inside the crowd is to attention on brand positioning. Your customers want to recognize you for who you are—the form of merchandise you sell, the first-rate of customer support you deliver, and the way reliable you are as a brand.

Chapter 11: Technical Seo Checklist

Why Technical Factors Matter in SEO?

Before we get into the details regarding the technical Factors We must first to know the concept of Technical SEO before we look at the steps to follow. Consider the following as:

Entry ticket: It's not possible to declare "I will just improve my website's technical factors and get it to rank higher within Search Engines. It's impossible. The technical aspect of SEO is only one part of the overall SEO.

Not essential, but important Consider this as "All you need to watch a movie at the theater is a movie ticket". If you're lucky, you can be on the 1st SERP page if you have a great domain name keyword, but only. If the targeted keyword is in addition to you Domain Name which is your Brand Name, then it's possible to be higher ranked.

Good Content + Technical Factors + SEO rankings: When you've got good technical factors and excellent Content, you can be sure of the top ranking on SERP (with keywords that are not competitive) and without Off-Page optimization.

Keywords - Decline In Importance The importance of keywords is declining each day. This isn't due to an excellent keyword or a poor-quality keywords used in your content. Rather, it's due to keyword stuffing in your website. You must adhere to Search Engine Guidelines, or otherwise! This means that the ideal keyword density should be between 1-2 percent. This means that the desired keyword is displayed between 1 and 2-times per word. If this is the case, the keyword is displayed enough to inform Search Engines what the page is about, but without keyword filler.

Therefore, it's not just about the importance of keywords but as well concerning keyword stuffing a problem which is becoming more crucial to Search Engines. Try to make use of synonyms for keywords to prevent Keyword Stuffing.

The next aspect you must know about technical Factors are as follows:

Optimize Topics and Improve Readability: The next factor to be aware of is "Optimized Topics" and "Improved Readability" are key aspects of Technical SEO. Utilize no cost online software to assess and improve the readability of your content prior to submitting your website on Search Engines or run any advertising.

H-Tags and Meta Description: The significance of H-Tags, Meta Description as well as other elements are extremely important in today's world. In the Search Engine Results Page, the searcher is able to read the "Title" and the "Description" of the listing, and decides to click the listing if it is in line with the query. Better tiles and a well-written description are essential. The

SERP listing begins with a "Title Tag or H-Tags" + "Meta Description" + (sometimes websites too)

In the past we would send these kinds of hyperlinks to Google however, now it is performed automatically.

Domain Authority + URL Authority Everyone knows that content is the king. Don't think that my post is too long, and it could affect my to rank. It is impossible to gain visitors without quality content. The algorithms of search engines are growing more sophisticated than human beings. It is true that Domain as well as URL Authority matter a lot. In the event that the exact content content was posted at website1, with Domain Authority of 20 and website2 has DA of 90, the outcome will favor the website that has a higher Domain and URL Authority. Google ranks trusted websites more highly than other websites.

Larger Documents and Slower Loading Make sure you don't use too many photos on your website. A blog that is larger is believed to be one with more than three images, and it may also have more than three ads. It is essential to test your blog and determine whether your blog is faster or heavier.

Google will give priority to websites with speedy loading and it's now becoming crucial to have speed. The slow load time isn't an ideal thing. Google prefers websites that load within less than five seconds.

As an SEO Expert, you must to be aware of the speed of your site, its title, meta description, the contents, etc.

Importance of Tags as Technical Factors in SEO:

Tags: We'll start with Tags since they are among the most important factors and why the majority of people mess with these elements in such a way. Don't make this trap and be sure to make sure you take Tags very seriously in your technical SEO.

If you are focusing on Blogging, you'll be better off using WordPress because there are many great plugins that will aid you in implementing Tags. It is not necessary to think about it in order to fill in the appropriate fields. If your site is custom-designed, you should learn how to apply HTML Tags on your site.

How Important Tags are for SEO?

Title Tags

Image Alt Attributes

Meta Description Tags

NoFollow / DoFollow Attributes

Heading Tags (H1 - H6)

Robots Meta Tags

Italic/Bold Tags

Social Media Meta Tags

Meta Keywords Tags

Viewport Meta Tags

The tags listed that are listed above should be your primary Priority. I'm not going to overload you with a long listing of Tags. You can read about the other tags here:

https://docs.google.com/document/d/1uEp weBili_gC7HQdMLzqNx4Ji4alS9uo4sCGnRyc CBI/edit?usp=sharing

Title Tags: When you visit search results pages, the primary thing that the user will look at the page's title. The page's title is vitally crucial. Title Tag remains SEO's first priority. The title is among the most important SEO elements, but it is important to consider other factors too. Be sure to give an original title for each page. Be sure to not exceed 55 characters. Include keywords in your title, which makes it more useful and easy for those who are looking for your name. If you own a brand name, make sure to add it to your title as often as possible since people are searching for it also.

Meta Description Tags This tag is crucial because when a searcher comes across your listing in a search engine and reads the title and description first before deciding they want to click on your listing, or click on it not. Be sure to not exceed 150-160 characters in the Meta Description.

If you did not add a Meta Description to your page and your page popped in the search results, Google will display the first 150 characters of your website. It's crucial not to put your keywords into your Meta Tag Description.

Tip: If you've got blogs with 500 characters do not copy and paste your blog to your Meta Description. Search Engines don't like it. Crawlers are smarter , and they know when to stop.

Heading Tags (H1 H1 - H6) First line of your page must at all times be an H1 Tag Heading. Start any page and go towards the top line (make sure you right-click the mainline, not anywhere else within the web page) then click right-click, and choose the option to inspect Element as well as "CTL+SHIFT+I". You will see that the mainline is marked with an H1-Tag.

Your page should include a Writing Design Structure , with H1, H2, H3 and so on. tags. Your website should be designed using H1, H2 as well as H3 (the most crucial headings) in the back of your head. However, your site should at the very least be labeled with an H1-Tag. When you use WordPress the title will be the H1-Tag automatically.

It's suggested to utilize more than just an H1-Tag. Also, it's recommended to have your blog with a design that's H1> H2 > H3, etc. If you require more, it is recommended to utilize Bold or Italic tags. This means that if you're looking to add additional headings, for example H4,5,6, you should include them Bold and Italic, too.

TIP The the Title Tag, Meta Description tag, and Heading tags (H1-H6) are the primary tags Search Engine Crawler looks for.

While the Title occupies one line of the SERP listing, Meta Description occupies 3 lines in that listing.

Bold/Italic Tags: You apply these tags immediately following H1,2,3 tags. Also instead of H4,5 and 6 tags, you can use Italic and Bold tags.

Meta Keywords Tags A long time in the past, Search Engines used to consider Meta Keywords Tags. But they don't anymore. Don't stress about it too much. Take a look at some of the pages' code, and you will see the meta name of their page is not filled with words: .

The Image Alt Attributes: Keep in mind that Crawlers (Googlebot) do not understand images since they do not know the meaning of an image is , but they do read the "Alt Attribute". It is therefore crucial to include an Alt Attribute to all of your images.

NoFollow attribute: Websites provide Link Juice to each other through hyperlinks. When you click on Website A, it takes you to Website B. and it redirects you to Website B.

If Website A contains the DoFollow hyperlink inside its HTML code If it does, then the Search Engine understands that Website A is providing Link Juice to website B. This implies that Website A tells Google that it is trusting Website B and doesn't wish to give or pass on Link Juice to it.

If Website A has the NoFollow link within its HTML code The Search Engine understands that Website A isn't giving the Website with Link Juice. This means Website A's message is to Google that it isn't a trusted source for Website B and does not want to offer it anything Link Juice. This is why Website A won't suggest to Google to follow the link to index".

Let's say that I have a website with 2 DoFollow links as well as a NoFollow link. If GoogleBot or another crawler browses my website's content it will also crawl the two links that have DoFollow however it will not crawl those with the NoFollow link.

"Robots Meta Tag": It isn't an issue with what's in the Robot.txt file. It's about the meta tag. that is in the text file itself.

This meta name has details about specific pages, including what they should index, what to not index, the best time to crawl, the number of times to crawl, and so on. For instance, if you wish to block a certain web page from search results, you instruct the crawler not to index it.

Social Media Meta Tags: Recently, Facebook introduced "Open Graph" to allow you to control the way your page will appear like when it is shared on Social Media. In this case, you could put Social Media Meta Tag HTML code to your page to inform Facebook that this page is one I am sharing, I would like you to add another title than the one this page uses, for example.

Viewport Meta Tag. It enables users to set how pages are adjusted and displayed across any devices. The issue is the concept of responsibility.

Brand Name Factor - Domain SEO Visibility:

We'll discuss "Brand Factor" and will review some SEO-related facts from the past. For instance, in 2016 Meta Description was not as important. The algorithms have been updated more frequently since then.

The year before, Webmasters started to use heading tags on their websites. Especially H1,2,3.

It is when the Brand Name Factor starts to play a significant part in SEO. We began to see more Brand Names as #1 on the SERP. Why? Because brands that are well-known often get ranked on the first page or in the top spots This means that brands and keywords that are associated with the brand are also influencing the results for general keywords. Domain names that have high Search Engine Optimization score better rankings for their URLs because of their trust with the Search Engines as well as due to their credibility.

Therefore, the Brand Factor depends on Your Domain Authority, your SEO Performance, and how well your website is performing as well as Your Bounce-Back Rate (BBR is the percentage of people who visit the website but leave without having spent a period of time that is considered important to the Search Engine), etc.

To summarize: The first listing on SERP will always be different from the rest of SERP. The first result could be directly connected to the brand's name. To understand how to understand the Brand Factor, you need to comprehend Domain SEO The visibility.

Domain SEO visibility: Let's look for the phrase "best golf club" 1. The first result and the 2nd URLs contain the word 'golf' as a key phrase. The content of these two sites isn't that impressive in comparison to the rest of the results. One of the 2 websites is www.golf.com The reason is due to the domain's SEO The visibility!

Visit any SEO tool and verify the URLs for your brand's SEO Performance. Example: https://suite.searchmetrics.com/en/researc h/domains/organic => Enter the URL (golf.com) => The higher the score the better.

There are instances when an online site has great content and it's optimized for search engines, however, it's still less ranked than the other websites due to its URL The Brand Factor.

Tips: When searching online, don't look following the pages at the top of the SERP that have a URL of the brand's name. They may not be at the top of their page in terms of content, however they could be more prominent due to their URL.

=> CONCLUSION: BRAND NAME DOMAINS ARE Good Factors for SEO Visibility.

Now the question is: Should you purchase the domain name with the keyword"?. We'll try to address this question.

More on Keywords and URL Parameters:

Do not forget to check out this article. It's an excellent guide that's called "9 Step ON-Page SEO Guide.pdf":

https://docs.google.com/document/d/18Dx 3dfSxJaMwDG90GTil3mjleKD39APbXNPu9q Rz1ls/edit?usp=sharing

For a summary of the post: If optimizing your site, pay attention to these crucial aspects The following are important: The page's "Keywords, Page Title, Meta Description, URL, Heading Tags, Weekly-based Page Content, Call To Action, Internal Links, and Images".

The importance of having a key word within the domain name has been reduced further, and in terms of ranking it has lost some of its efficacy. It's obvious that the algorithm for searching is now less important to the keywords within the domain name.

In the beginning, we must be aware that there are three kinds of domains.

EMD also known as Exact Match Domain: We have discussed this in the past. Example:

www.googlekeywordtool.com used to rank #1 higher on the SERP for this match. But not anymore. Exact Match Domains do not offer the same results they did previously.

A New Google Algorithm: Even with EMD there is still a need for good content, excellent quality backlinks, and a good authority on your site.

There is a reason why Google has only one page on the SERP that has EMD Therefore don't be concerned when you don't see an EMD for your domain. There is an undisputed fact that having a key word within your domain is a good idea.

The following tale is the reason Google does not like EMD no more. It used to be a very good strategy back then:

In 2012 and 2013 it was it was easier to rank a site. If a film is set to release, someone designs a review website using this name and it gets ranked effortlessly. => Example: www.Movie1Review.com. It receives a significant amount of people visiting it for about a month but then it slowly dies. Then they develop a brand new website whenever the next movie comes out and they name the site www.movie2Review.com and then they give it lots of Juice Links to Movie1Review.com and the like. Today, Googlebot got smarter and more sophisticated and is now thinking primarily about quality content. Don't be concerned about your domain's name being not branded.

PMD, also known as Partially Matched Domain. Try to focus on 3 or 4 keywords to build the domain name.

Brandable Domains: Many of the most popular websites on the internet come with a brandable domain. Example: Amazon, Uber, etc. The keyword is an established brand. They're not focusing on the keyword but are focused upon the Brand.

Be sure you concentrate on the most Brandable Keywords if possible.

It's great to have a keyword , but you must focus on branding the Keyword by itself. Consider going beyond Keyword and think "Brand Keyword".

Is dot com .com useful or not?

There is a lot of talk about this: the extension of your domain's name does not have any significance any more. It's all about user experience. They can consider ".com" rather than ". somethingElse". But, if you search Google, 99.99% chances that the first search result is concerning ".coms". Are these coincidences or an indication from Google? There is no way to know.

Tip: If your goal for your domain name to make a band name and you want to create a Band Name, don't use any other domain extensions besides .com since you'll spend a lot of money and time to create that brand and eventually, someone will purchase the .com to take on your.

URL Length Parameters:

If you take a look at the SERP with the highest rank you'll see that the median number of characters found in the top 10 URLs is about 55. We're not talking here about the domain; we're talking about the entire URL (www.site.com/your-link-here).

TIP: Stay away from these types of links: "www.site.com/12cv?sadf-0238etc" => Hard to read and they look spammy. If it's an e-commerce website, Google may know that you need to keep track of the ID of your product, however when you're blogger Google doesn't really care about it and penalize you.

HTTPS + SSL = HTTPS?

As you are aware that in July of 2018 the Google Chrome browser began to mark websites that do not have HTTPS with the tag "Not Secure". If your website isn't equipped with an SSL certificate the site will be displayed with the tag "Not secure" in red color, which the user is not to overlook.

SSL Certificates help ensure that your connections are secure and secure. If you do not have an SSL Certificate, you have an unsecure connection and nobody will be able to be able to trust your business with you since the transfer of data between the website and the user isn't encrypted. If you are an SEO professional You must focus on this as early as you can.

STEP 1: INSTALL GOOGLE ANALYTICS

5 SEO Metrics to Measure SEO Performance:

We previously covered the basics of SEO in detail, and today we'll look at the methods to gauge SEO performance by taking a look at five different metrics using two free tools: Google Analytics and Ahrefs tools. These 5 Metrics are Website Traffic, Conversions, Revenue, Keyword Ranking, & Link Building. These five metrics will let you know how SEO success could look like and how it is evaluated. They will tell you how many people came to your site from search results, and whether you've had the desired results.

Google Analytics Tool: You need to install Google Analytics Tool: https://analytics.google.com/analytics/web

This tool provides you with detailed information about visitors' numbers and information about search engines. Apart from the tracking of your website's visitors, Google Analytics will also provide you with insights regarding the terms that visitors use to find your website. One of the top FREE training courses available on Google Analytics:

https://analytics.google.com/analytics/academy/

Metric 1. Website traffic: It's crucial to gauge your website's traffic as an important SEO metric. This includes the traffic to your website that is being generated by Search Engines. It's easy to see that Google Analytics is a simple way to see the amount of Search Engines' traffic is making an impact on your website.

Measure 2: Conversions = This is the second metric you must consider to gauge the effectiveness of SEO. We want to determine how much traffic Search Engines are bringing us but also know the percentage of that traffic is being converted (did you achieve your goal? And did anyone buy something or users sign up to our email mailing lists?). You can examine the results of all these in the Google Analytics Section called "Goal Completions & Conversion Rate" on the dashboard for Google Analytics.

Metric 3: Revenue The final measure that you could examine using Google Analytics to measure success is Revenue. When you've got a payment feature on your site and have successfully connected that process for Google Analytics, you can determine the amount of dollars that comes to your site via Search Engines.

II.Ahrefs tool: The main feature of this tool is to determine the position of your website's search results for the specific Keyword as well as the number of Backlinks your website has accumulated. Ahrefs is a premium tool but you can use the free version: https://ahrefs.com/keyword-rank-checker

Ahrefs tool is one of the most highly recommended SEO tool for Bloggers and every website owner. It is a must-have tool if you wish to rank on Google. The tool can help you not only increase the number of visitors to your site, but aid in keeping in mind your competitors' search engine rankings, from where your competitor receive their backlinks and what amount of link juice is your competitor receiving from a particular backlink. It will also help you understand which keywords are in the news.

Metric 4: Keyword Ranking Keyword Ranking: Another great measure to examine. It's about how you rank for specific keywords. On Ahrefs scroll until"Organic Search" and scroll down to the "Organic Search" section to discover the "Top 5 Organic Keywords". They are the position on the search for these keywords as well as their ranking in the search results. Click "View Full Report" to learn more about this.

Metric 5 Backlinks Go down Ahrefs and then to"Backlinks" and then to the "Backlinks Profile" section. There, you can look up the number of backlinks , which is the number of websites that are directed to your site. It is thought of as an important metric to consider Do you have a steady increase in your backlinks over time ? This can help build your authority and allow you to get higher rankings in results of searches. If your site is brand new and you do not have any backlink information:

Keyword Ranking and Backlink Metrics are excellent key metrics from Ahrefs. Ahrefs tool. If you're just beginning with a new site It is not advised to upgrade Ahrefs due to the fact that a large amount of Metrics information won't be available due to no traffic to your site as of yet.

Visit www.ahrefs.com then type in your URL for your website => You will be able to download the report for each of the sections:

- Dashboard - Alerts - Site Explorer - Content Explorer - Keywords Explorer - Rank Tracker - Site Audit - Etc.

The left sidebar contains many useful elements you can choose from.

To summarize: We've discussed the 5 most important metrics to gauge your success by using two free tools: Google Analytics and Ahrefs.

In the next step, we'll discuss how to improve your website's performance for Search Engines and get real results.

STEP 2: OPTIMIZE AS MANY TECHNICAL FACTORS AS YOU CAN

How to Optimize Your Homepage:

In the previous section, you learned about how to evaluate success and I taught you the metrics and tools to concentrate on. In this section you will discover the best ways to improve your homepage to be optimized for Search Engines.

One of the most crucial factors you need to consider on your site is when someone type in your company's name or business name into Google you want your website's home page to appear attractive and reflect your company. You must have a great description of your company here. It also entices visitors to click.

Optimizing your homepage won't get you lots of traffic from Search Engines because there's not any people that are aware of your site. So, if you want to increase your traffic it's a good idea to set this up as soon as you're just getting started. It's easy to show up on the first page of search results for your company's name. But, we're not just optimizing our brand's image however, we'll discover the keywords our clients are searching for on the internet, as that's how they learn of our company.

Let's make our website more technical. Factors

When you write your blog, you must ensure that you are taking care of these factors listed below:

How to "Optimize Writing & Topics" to "Improve Readability". => Done

How to "Use H Tags & Description" Properly. => Done

How do I know what "Domain & URL Authority" is. => Done

How do you deal with "Larger Documents & Slower Loading" It's not done yet. We will discuss this aspect in STEP 4 "Improve Website Performance & Speed"

"You don't have to be a Web Developer to implement those elements on WordPress but you need to have a good amount of HTML coding experience to do this on HTML pages."

Sometimes, it is difficult to deal with these aspects in your head because the editor isn't as powerful. If you want to install a plugin dubbed "Use any Font"

Once you have installed"Use Any Font" plugin, after you have installed "Use any Font" plugin then go back to the editing mode, and you will notice that there is an icon "Any Font" where you have more options for fonts and paragraphs. This plugin, or other similar plugins, will make writing more efficient.

Also, you must install a different plugin known as "Yoast SEO" by Team Yoast (Make sure that you install the genuine one and not any duplicates). This will assist you to improve your SEO. We'll cover Yoast SEO in more detail later on in this article.

With these two plugins, you can improve your content as well as SEO.

If your website isn't a WordPress then you need to make use of HTML codes to improve your H-Tags and other.

It is important to design all your content using one font. Choose all your text and then choose The font "Verdana" size 12 is the typical.

Next Step: Install This Google Chrome Extension called "Grammarly". It will assist you in improving your writing with English spelling, grammar English grammar, spelling.
https://chrome.google.com/webstore/detail/grammarly-for-chrome/kbfnbcaeplbcioakkpcpgfkobkghlhen?hl=en

Here's the format you should adhere to when creating your blog or post:

"URL -> Title" -> -> "Body -> Meta Description" -> -> "Images".

While working on the post, make sure that your URL (permalink: www.site.com/yourSlug) is no more than 155 characters total.

Once you have installed Yoast SEO, after you have installed the Yoast SEO plugin and are writing your article take note of the SEO score and the suggestions on the right side and make any necessary adjustments as you proceed. Keep in mind that readability is heavily dependent on the niche. Be aware of Grammarly when you are reading also.

Most importantly, you must to create your H1,2 in a correct manner. For now the title is H1 and you must make H2 and H3. If you are on an eCommerce website Your H1 should be the Product Name.

For an H2 tag you need to do this manually. Choose a heading in your text that you wish to convert H2 and then change it the heading from Paragraph in H2 make it a size of 14pt plus bold it.

When you think the term "Paragraph", select the text you would like to make an article, then size it to 12pt, then create it into an article.

Then: This page is read from H1 through H2 until Paragraph.

This is also important: Try to add categories to your site whenever is possible => You are telling Google that my site is categorized/organized. Google loves it!

Next, you must focus on Tags. This is the field that appears located on the right-hand left side of your WordPress page that you can choose to insert tags (keywords). The purpose is to include the keywords that you've included in your blog post, into your

Tags field. If you're working in HTML you can add meta tags to your HTML code.

When you are working on your blog or page Pay attention to the 3 SEO tabs at the lower right of the page you're working on:

1. "Readability: To understand the readability of the post"

2. "Your Focus Keywords: To understand how you are using your focused keywords in the title and meta tags",

3. "Add Keywords": It's not free and is an upgrade. You can click on one of these tabs to view the report, comments and suggestions of SEO Yoast and follow from there.

Let's go over Yoast SEO in more details:

I'm assuming you've installed the Yoast SEO plugin:

This WordPress plugin will allow you to concentrate on your keywords when creating your blog article. It also lets you include Meta descriptions, titles, tags and more. It also comes with the "Page Analysis" Feature that will help you improve your website's content and improve its SEO.

To learn more about this technique you can type your domain's name in the Google search engine. Google. The result what you'll end up seeing are pages from your site that Google has spotted:

I can see: The URL for the webpage and it's title and the description for the web page. This is the area we'd like to improve.

Let's now optimize it. It's easy to do using WordPress websites. The great thing about "Yoast SEO" is that it can help you with numerous things in motion beginning with this specific task: optimizing your description and title.

After downloading the plugin, go to the page you wish to improve. The goal is to enhance the title and meta description. The best way to write an effective headline is that you include the company's name, along with some keywords which represent your company. You can check this great example by typing "backlinko" in the search bar on Google. Look at their title on this page. It's their brand name plus the details of what their business is about. The title of their business is: "Backlinko: SEO Training and Link Building Strategies".

The second thing you should include is your Meta Description and this is only a small portion of your site that is displayed in the results of a search. The reason is that Google generally will pull the Meta Description automatically from the page, however you can make this happen using "Yoast SEO" to better match your requirements. It's strongly advised to begin in your meta description with the word "verb. It is essential to provide your readers with the followinginformation:

What is the outcome for them?

What will they do?

What advantages the product/service can bring to them.

Examples: "Learn a ..." beginning with an adjective.

In the end, I'd suggest that you examine your competition and look at other blogs or platforms and then try to imitate them.

One last point to remember is that once you've done this, don't be expecting that it will appear in results of a search immediately. Google will check your site every few days for up to a week's maximum to determine if anything has changed.

Now that you have a clear idea of the best way to present your description and title Let's think about what you would like your listing to appear on the SERP:

LET'S START WITH "FOCUS KEYWORD" TAB:

The next step is to include "Focus Keyword": The Focus Keyword is a field in which you can enter keywords or phrases. After you have entered it the plugin analyzes and determines if the specific keyword is inside the SEO Title the H1 tag, body as well as in other Elements.

When "SEO Yoast" is not satisfied with your Title and Meta Description It will offer you suggestions.

Then, you'll need to finish the Next, you need to work on the Featured Image of your blog.

In the end, as a result of these SEO steps, you will receive the following report at end of the page. It is a report that contains recommendations on how your blog is doing or not doing depending on the titles, links and meta tags, for example. It's an excellent report: follow all the recommendations. For example, if a plugin recommends that you improve the internal links of your site, this implies that you must internal connect your posts and pages by using keywords in your articles. Another example In the event that you don't have the keyword you want to target in your Meta Description, or does not appear in your first paragraph the keyword will be flagged in the report. This report also show your exact-match keyword density as well as offers suggestions. The report also goes through your photos and give you a flag if there's missing images featured in the feature or alt attributes. => USE THIS REPORT TO YOUR ADVANTAGE!

LET'S WORK ON "READABILITY" TAB:

Select the Readability tab> Read the issue and then fix the issues. Many of them are simple solutions.

You should go over the text in relation to the way a 12-year-old child would read the text. A score of 60 or more is considered to be good.

To run a Readability test, go to http://www.webpagefx.com/tools/read-able => Quick and easy way to test the readability of your work.

Check the positioning of your AdSense links on your page. Don't add more than 3 Links per page.

Be sure to include at minimum 100 words between two H2 tags. This will help you to pass your Readability Test.

Verify that your links aren't broken.

Maintain a balanced space separation between text and images.

Infographics (with statistical data) are a good thing to include. They're great to be seen by Google's search engine. They aid in ranking your website more highly, as they strengthen your backlinks.

If you write blogs about product comparisons, you'll should make use of a plugin for comparison of products. I recommend you conduct the Google search to find the most effective WordPress plugin for comparing products'.

To recap The most essential elements you should consider in creating your blog or article, however we didn't cover Page Speed Optimization. This is something we'll cover in step 4.

It's a simple method of optimizing your company's brand name for search results. In the following step we'll be by ticking these points off our SEO checklist for technical SEO. We'll also include and verify your web site to Google's Google Search Console.

STEP 3: ADD & VERIFY YOUR WEBSITE IN THE GOOGLE SEARCH CONSOLE AND SUBMIT YOUR SITEMAP FILE TO GOOGLE

In the previous step, you reviewed how to measure your SEO's success, and optimized your website for search results. In this step, you'll go over Step 3 of the 25 points checklist: Add and verify your site in the

Google search console: www.google.com/webmasters and submit your Sitemap file to Google. This is a second element of Technical SEO.

Introduction to Google Search Console:

Google Search Console or GSC is a set of tools and resources to assist Webmasters, Marketers and SEO professionals to track the performance of their websites on Google Search Console. Google Search Index. In fact, Bing, Yahoo, Yandex in Russia as well as Baidu within China have Webmasters and Web Search Console instruments. These tools let you assess how the Search Engines view your site by providing information on the status of indexing as well as penalties, and crawl errors which you may be subject to. They also permit users to send their Sitemap file to these consoles, and more.

To access Google Search Console, you must connect your home (your domain name) in the console. You must confirm that you are the owner of the domain and upload your Sitemap file to Google Indexing, through the GSC service. The Sitemap file contains all of the links you would like Googlebot to search and index Search Engine Results.

In the first instance, if you hear someone talking about Google Webmasters, they might be talking about Google Search Console. It is important to know that each Search Engine has its Webmasters. Both tools are equipped with the identical functions, with the exception the fact that Google Webmasters is the old version, and Google Search Console is the latest version.

As of today, Google completely mover to Google Search Console: https://search.google.com/search-console/welcome

How to Add and Verify Your Website in The Google Search Console:

Log in to the Google Search Console =>
Select your Property or add another. It is
necessary to confirm that you own the site.
Choose to download the HTML verification
file and then upload it to the root directory
of your computer employing FTP software
such as FileZilla and then return in Google
Search Console GSC and click on verify.

Very important When using WordPress then
click in"Step 3 "Alternate Methods" tab on
GSC Select one of the four options that we
want to select"HTML Tag "HTML tag" option
and take the meta tag's code. You can go to
"Yoast SEO" => General > Click
at"Webmaster Tools" => Click on
"Webmaster tools" tab =After that, you
copy the code in your "Google Verification
Code" field. Do this: only need to enter the
long serial number after the content=" and
without quotes".After that, => Click Save.
Return on your Google Search Console and
click the "Verify" button. The website has
now verified and has been added to.

To summarize:

You've already added your website to Google Search Console. However, you haven't uploaded your sitemap file. This means that your website won't appear in SERP until after you have invited Googlebot to browse your website (without inviting Googlebot it takes longer for your site to be listed on SERP. Googlebot will visit your website at some point, but without invitation, it'll require more time.)

Through Google Search Console With Google Search Console, you can view reports of indexing status of all pages that Google discovered on your website. It also shows what Search Queries that you're appearing in the results of a search. Also, you can see the crawl errors on the pages you have and look up any potential penalties for your website. This is a great tool for technical purposes. After you have your Google Search Console installed and it is able to start gathering information.

You also want to check this guide on Search Console Help: https://support.google.com/webmasters/#topic=9128571 which explains it in a lot more detail.

After you've added to your site and confirmed it's a valid property you'll find that you don't have any data on file. It takes time before the Search Console can collect data and analyze it for your website. Given that you've no information yet is the main thing you need to do now is to set it up and let it begin to collect data. And then, later, you will return to check what your Search Engine views your site with the help of reports on indexing status as well as crawl errors and penalties and more.

After that, you'll be able return to your dashboard

Make sure you check your Search Appearance for such things as Structured Data or Rich Cards and so on.

Make sure you are checking your search traffic.

Take a look at the links on your website.

Take a look at your internal Links.

Take a look at some Indexing reports listed under "Google Index".

Check out some Crawling reports listed under"the "Crawl" section.

Etc.

Don't be overwhelmed by GSC and focus only on these important aspects. These are the most crucial Google Search Console Features:

"Performance" => This tool will be your dashboard from which you can get an overview of your total clicks "Total impressions," and any other information related to performance. Spend an hour or so analyzing this section.

"Coverage": This a"Coverage" report is a Console Coverage report which provides details about which pages on your site have been indexed. There is also the list of URLs that had difficulties when Googlebot attempted at crawling and indexing them.

"Sitemaps" => Displays the websites you've submitted for inclusion at the top of Google Search and allows you to add additional sitemaps should you wish to.

"Search Analytics" => This is the amount of clicks you're receiving through Google to find your website. If you use this option, Google can assist you by recommending keywords that you can include to your blog post to make it increase its visibility. With this Analyticsfeature, Google will inform you about the terms that users entered in the Google search results and that have led to your site's display in the SERP.

For instance, you notice that your website is ranked #11 when people search for the keyword 1. However, there isn't an essay or on the specific keyword. This is a great gain from Google and an amazing fact that Google recommends you to take action. What you should do is write a new post or create a new section that contains the keyword1 and you'll find your website appearing EVEN higher in the SERP.

"Messages": This is the place where notifications from Google are made available. For instance in the event that Google finds "Negative Hat" activity on your website and it notifies you via a message to ensure that you are aware of this.

"Search Appearance" => This describes everything about the way your website is displayed when you search on Google Search Result desktop and mobile. It includes "Structured data, Rich Cards, Data Highlighter, HTML Improvements, and Accelerated Mobile Pages".

"Search Traffic" => This is a reference to Your "Search Analytics, Links to Your Website, Internal Links, Manual Actions, International Targeting, Mobile Usability".

"Google Index" => This includes "Index Status, Blocked Resources, Remove URLs".

"Crawl" => This is a crucial section. It contains the following:

.Crawl errors,

.Crawl stats,

.Fetch as Google,

.Robots txt Tester,

.Sitemaps, and

.URL Parameters".

After you have registered and verified your website on Google Search Console, once you have added your website to Google Search Console and have an understanding of the key functions that are available in

Google Search Console, you must upload Your Sitemap data to Google.

Create an Sitemap document for the website:

In order for your site to begin appearing in Google Search Engine, there are three steps to follow: Create a Sitemap • Add your website to the Console => Invite bots to browse your website.

Create an Sitemap for your website .It's what you're working on right now.

Include your website in Google Search Console => You've already registered your property and verified it.

Invite Googlebot which is also known as Google Crawler to visit your website , and it will eventually show on Google Search Results. We'll discuss this further in the section titled "Invite Googlebot and submit your Sitemap file to the Google Search Console".

A Sitemap is comprised of all the Pages that you would like for submission to Google and be displayed in the search result.

There are two methods to create a Sitemap

Method 1 - WordPress: You use a plugin. Visit the WP Dashboard. If you're not yet installing "Yoast SEO", it's time to get it installed. Once it's installed, click"Features". Click on "Features" tab on top. Then scroll down to search for "XML Sitemaps" and then enable it.

The plugin generates a few sitemaps files for your site:

A sitemap file to help you organize your posts, another to manage your content, a different sitemap for pages a different category one, a third one for authors/writers, etc. When you browse the sitemap of posts, you'll see the complete list of your posts. If you go to the sitemap for pages, you'll see the list of all your pages and so on.

Very Important: You should not submit your post tags, your category as well as your author sitemaps to Search Engines. SEND ONLY Posts and Pages Sitemap Files.

Method 2 - Non-WordPress If you don't use WordPress visit www.xml-sitemaps.com and create your invite to Googlebot and upload the Sitemap File to Google Search Console file there.

A good example of an XML Sitemap format is: http://www.sitemaps.org/protocol.html

A good example of an HTML sitemap format is: http://www.apple.com/sitemap

After you have created your sitemap it is time to add it to your webserver's root directory.

It is possible to block irrelevant links on Your Sitemap files from Search Engine After you have an inventory of all your links in the Sitemap You may wish to prevent certain documents from being displayed in the Search Engine with the tags known as NoFollow that were previously discussed.

Once you've got your Sitemap document in hand and ready to go, you need to send it to Google Search Console as an invitation to Googlebot or, also known as "Google Crawler", to browse your website and index your site to search for Google Search Result.

Invite Googlebot and upload Your Sitemap File for submission to Google Search Console:

In this stage you have added and verified your site to Google Search Console and you have created a sitemap for your site. The next step is welcoming Googlebot as well as Google Crawler to browse your site and get it listed in Google Search.

Visit the Google Search Console account and then click on your property. Then click on Crawl, then click on Sitemaps: You'll notice that you don't have any sitemaps. You have the option to create one Click"Add/Test Sitemap" button "Add/Test Sitemap" button: Keep in mind that we do not want to share all sitemap files and we only want "Posts and Pages Sitemaps" You can copy only the sitemaps that correspond to it: post-sitemap.xml and add one by one , and then refresh following each submission.

Important point: click on the "Test" button first before you click Submit. This way, you can test your sitemap for any errors, and then you correct the issues prior to submitting. VERY IMPORTANT!

After refreshing, Google confirms with you that you've submitted the post-sitemap.xml or page-sitemap.xml and provides the number of pages, the number of images, etc.

Very Important: Until the present, you've sent your sitemap to Google and Google confirmed that they have received your request. This means that Googlebot received an invitation from you to visit your site and then index them. Once Google accepts your invitation, it will take about a week to index your pages.

Then, you can follow the same process with the second sitemap, page-sitemap.xml. Remember to Test and Refresh.

Adding Website and Sitemap to Bing Webmaster.

It is also essential to include your site's URL and upload your sitemap file in Bing Webmaster as well.

Visit www.bing.com then look up Bing Webmaster. www.bing.com/toolbox > Register for your first account. Create your Website by filling in some details about your webmaster and the website as the webmaster. You will need to add your two regular sitemap files You are able to skip any optional information by clicking Save.

In this way you can copy the meta tag's code supplied by Bing and put it on your site's tag. If you are using WordPress use, visit Yoast SEO > General Webmaster Tools Tab Copy the code into the Bing Verification code field. The only thing you need to do is paste the long serial number after the contentis" and without quotes". => Save.

Another option is to download smaller XML document from Bing and upload it to the root directory on your server (basically you must follow the instructions there).

The procedure is exactly similar to Google Search Console and Yahoo Verify the website and upload Your Sitemap file. Once you've completed the process, return to Bing and select "Verify".

You can also add additional sitemaps like we did using Google.

How to Submit Blog Posts to Google For Fast Indexing:

We'll go over the steps to submit your blog or website to Google for indexing speedy. If your Sitemap is for every blog or page you have had have submitted to Google and you're not sure the exact date when Googlebot will come to your website and then crawl your site. For instance, let's take that you are in the middle of a News which just appeared and you'd like to be to the top of SERP before everyone else does and you are unable to afford to sit around waiting. Therefore, you should upload your site to Google in the shortest time possible.

Here's how you're planning to send your post to Google and fast:

Click on Google Console then click on Crawl, then Fetch using Google Now you are able to publish your blog. (basically, instead of submitting www.site.com, we are submitting www.site.com/blog.html) => Click on "Fetch" not on "Fetch and Render" => At this time, Google will give you an option to request Indexing. Simply hit"Request Indexing. "Request Indexing" Button =Choose "Crawl only this URL" and then click"OK. You're done. This means you have requested indexing of your article. This is the Desktop kind of submission. You can do the same for Mobile & submissions for smartphones too.

In the next section, we'll be discussing various versions of your site. It is crucial to know about SEO. Sometimes, you get audit errors regarding your website version. Search Engines are misinterpreted in between http://www.site.com or http://site.com = So, it is important to know which version you're transmitting to Google!

Adding Different Version of Website to Webmasters:

To simplify the concept it is important to inform Google the Property kind (www.site.com or http://site.com) you wish to showcase in the SERP.

Technically speaking, you'll need to eliminate this issue: That's right that you must have two types of websites' properties: http://site.com and www.site.com. So, determine which property is not present and then add it: Sign into Google Search Console => Select the Propertyand then note which type of property Google has. In the event that www.site.com appears to be the one Google created then add the http://site.com Property today. Simple as that!

Tips: The very first time you're adding a home to Google Search Console, it's typically added under site.com instead of www.site.com therefore it's best to include the http://www.site.com the property.

Visit the Google Search Console dashboardYou will notice that you have two properties that are the same.

Important When you have two properties that are of the same kind and you need to inform Google the website versions you would like to display in search results". To do this, click on the property you wish to use . click on "Settings" gear icon => Site Settings Select the version you wish to display on the SERP. Save. That's it.

Another option is to include "301-redirect" from the www version to the website's original. This requires the knowledge of HTML programming. Do not worry If you're not comfortable about how to accomplish it and simply go through the steps mentioned in the previous paragraph.

That's it! The process of registering your site in Google's Search Console can be easy. In the next SEO step we'll be testing the speed of your site. This is the 4th Step of the Technical SEO Checklist. We will discuss the Website-Page Speed and ways to improve it by using tools such as:

http://tools.pingdom.com or similar.

You could also test your website to see if it is speedy using Google Server:

https://developers.google.com/speed/page speed/insights

We'll discuss the various WordPress plugins that you can install.

We'll discuss the various "Compression Techniques" you can employ such as cache and CDN to enhance your distribution or delivery network.

We'll go over how to deal with images that are heavy and ways to make them more efficient.

STEP 4: IMPROVE WEBSITE PERFORMANCE & SPEED

How to Double Conversions by Improving Website Speed:

Page Speed: Following the advancement in smartphones Google will now be focusing its attention on page speed. It is important to check your Page Speed frequently and track the score of how quickly your website's content loads. So, if your site is SEO-friendly but has a slow loading speed, it will not show in the SERP.

Have you heard that website speed is the number one reason that people stop buying on the internet?

Take a look at the following case study

https://blog.radware.com/applicationdelivery/applicationaccelerationoptimization/2013/05/case-study-page-load-time-conversions/ - 2 seconds Improvement in Page Load Time can Double Your Conversions.

Follow the three steps below to improve the speed of your site. Then go on to determine if something isn't working or you'd like to update and correct:

Step A: Evaluate your website's speed using Google PageSpeed Insights https://developers.google.com/speed/page speed/insigh

Step 2: Download the following 3 free WordPress plugins to improve the speed of your work:

WP Revision Control This lets you free space e.g. just keep the most up-to-date five versions of each webpage on your site This is:

1. Upload and enable and activate the plug-in "WP Revisions Control" by "Eric Hitter".

2. Click Settings > Write, and choose the settings under the WP Revisions Control.

- WP Smushit resizes, optimizes and compresses all your images:

https://wordpress.org/plugins/wp-smushit

. Upload and then activate your plugin "Wp Smushit" by "WPMU DEV" and then select the plugin and click "Get Started".

- WP Super Cache - Uses this smart trick to load pages faster: https://wordpress.org/plugins/wp-super-cache

. Upload and then activate your extension "WP Super Cache" by "Automatic" to help you create static HTML pages.

Step C: Test your website's speed a second time using Google PageSpeed Insights (aim at 80or more):

https://developers.google.com/speed/pagespeed/insights

Notice: Hundreds of thousands of businesses utilize these plugins, but when they create problems you can delete them and then search for alternative solutions!

If your site is loading slow and displaying, you must focus on the speed of your website and then its content. It is possible to have great content, but if your website's speed is not optimal and you are not successful in SEO. You can download this Chrome Extension known as "Analyze Page Performance" from

https://chrome.google.com/webstore/detail/analyze-page-performance/hemibacgndhdhkfahkjdedjdgfapmfki?hl=en-US

It's exactly the same. This tool will help you assess and comprehend the speed of your site when you open it up in Chrome. Chrome Browser. It examines your page's performance.

After you've been working on improving your page's efficiency and speed, we'll look at additional details on this issue so you can take a look and make adjustments as you see the need.

What is the cause of a decrease in the Page Speed?

Start by reading this good article first: Google plans to give slow websites a new

"Badge of Shame" in Chrome => https://www.theverge.com/2019/11/11/20959865/google-chrome-slow-sites-badge-system-chrome-dev-summit-2019

It is crucial to look over Page Speed and the Optimization of your website". As you may have guessed, Google starts giving importance to page speed. Currently, it is being considered as an important ranking factor in the Google Algorithm.

What can slow the speed of your website? Before answering this question it is important to know the reasons why a site or page is slowing down. Below are a list of things that might be the root cause for a slow websites or pages in general . You need to look at every one of them.

Your host's address - Verify their

* Uptime/Downtime

* Security

* HTTPS/SSL

* Support

* SSD (Solid State Drive hosting - Faster) vs. HDD (Hard Disk Drive hosting - Slower)

For instance, you want to verify the configuration of your server to check if the site is hosted on SDD or HDD hosting. Also, look at your "RAM" you are using as well as what "Space" you are given and how big are your site's files. If your site has excessive images or numerous JavaScript file, you can cause your website to slow down. This is why your hosting platform is among the most crucial aspects! You must monitor it continuously.

External embedded media: A more heavy YouTube and Vimeo video could slow your website. They could reduce your loading time. In order to increase load speed you must upload the video to your server. JavaScript animation, or animated flash or any other external media like from Facebook could make your site slower.

Too many advertisements It is recommended to add at least 3 ads to optimize your website. It's recommended to have at least 2 texts and two banner ads. Ideally, 1 and 1.

Widgets: They include Social Sharing Buttons and Comments areas that could be detrimental to the speed of your website.

Unoptimized Browser, Plugins and apps: Think of those users who have browsers that aren't optimized, such as than explorer or chrome, or Firefox. Thus, do not make use of file types to upload images that won't be compatible with other browsers. Also, make sure you don't use too numerous plugins or programs. They'll make your website take longer to load in the future.

Large Images: Make sure you optimize and compress your photos. There are a variety of free online tools that can help you do this. I use this site to handle my majority of my pictures need: www.toolur.com.

Your Theme: You need to select a properly optimized and flexible theme. Make sure you choose an unpaid theme since some of them aren't optimised. Additionally, free themes carry the possibility of being easily hackable.

Incorrect Coding: If your HTML isn't properly formatted or if JavaScript is included in your header , or if your CSS is located in your footer, it can cause your website to be slower. The rule of thumb is: JavaScript should be loaded at the lower part on the webpage, and not within the header. If it is not done, it can create random blockages.

• Place Your JavaScript at the end of your code to ensure that users don't see blank pages while JavaScript starts loading.

=> A great tool to optimize your HTML code
=> W3C Validator: https://validator.w3.org

After you've improved you HTML code, you'll want you to improve your CSS code too. There are tools to help you do this. I recommend using: http://jigsaw.w3.org/css-validator

Why Page Speed is Important?

Once we know the reason our site might have a slow speed, we'll look through "why Page Speed is important" and then discuss "what are some reasons that website pages are slow" and "how this can affect Google ranking and Users as well".

Studies have shown that one second delay in the page's Load time could result in "11% Fewer Page View + 16% Decrease in Customer Satisfaction + 7% Loss in Conversion". The modern day user want to get results as quickly as they can. This shows that it's not just Google but Users also prefer faster websites. Slow websites tend to have a higher "Bounce Back Rate".

A recent report from Amazon in relation to Page Speed revealed: "How 1 Second Could Cost Amazon $1.6 Billion In Sales every year".   =>

https://www.fastcompany.com/1825005/how-one-second-could-cost-amazon-16-billion-sales => You need to keep the focus on how Page Speed is beneficial and how it can affect Users, Google ranking, as well as your website.

As an SEO professional As an SEO Expert, you must have an understanding of the speed-related solutions and how it is possible to improve your site speedier and more optimized in terms of speed, images, and so on.

Before we discuss ways to fix speed issues on your website Let's look at the speed test and outcomes of your site. Click here: https://gtmetrix.com to Enter your URL Be aware of the following elements:

PageSpeed Score more than 85% is considered good.

Fully loaded time Anytime you go over 3-4 seconds can be a problem. This is your primary concern. It is possible to optimize your page to decrease this score and this could be an issue typically because the page files are too large.

The Score for Requests is anything lower than 80 is fine.

You may also test your website on these platforms:

* Google's PageSpeed Insights" to get an official score: https://developers.google.com/speed/page speed/insights/?url=udemy.com

* Pingdom Website Speed Test:

https://tools.pingdom.com

Test the performance of a site:

https://webpagetest.org

* For Mobile To test, use:

https://www.thinkwithgoogle.com/feature/ testmysite owned by Google.

* Web Page Test:

https://www.webpagetest.org/compare

It is important to comprehend how "Client-Server" technology behind this: Client requests data from your Server. Your server responds by transferring those data back through your Data Center to the Client. The time for server response differs based on the Server and your web page.

Utilize "Caching" to make your page load speed increase. This means instead of waiting for your server respond to the request of the client and then send those dynamic pages directly from databases, can force the server to transmit static page (copy that of pages with dynamic content) out of the cache. Then you must understand how to create static files of your website with caching. This means that your load request isn't sent to the database with every page visit Instead the server makes use of pre-saved static pages.

The problem can persist even if your photos (who are static in default) are large. To resolve this issue it is necessary to improve the quality of your images.

Other possibilities include:

Certain ads section issues can cause slowdowns on your website.

The bottom portion of your site first, and then the top part of your website later can be an issue. It may show a blank screen and slow your site.

Tips: Everything on the internet depends on two points sending requests and receiving Response. Your task is to focus on two things that are: how to make a request faster and create a static file and then send the response FASTER.

You must understand and apply this idea. This is how Optimized websites function.

Let's recap: In step 4 you optimized your Page Speed Performance , and covered other aspects. Let's look at the outcomes.

Let's Get Started - Analyzing Website for Page Speed:

When working to improve Page Speed, always start with these two tests: These are the tools that are recommended:

http://www.gtmetrix.com/

https://developers.google.com/speed/page speed/insights

If your website is not performing as it should the two tools will say that the website is not performing well and provide suggestions. In addition, Google offers additional suggestions.

The image below is about the steps that you must follow in order to optimize your site's speed. We'll briefly go over each of them today and in the future, we'll take on the whole page speed Optimization SEO process.

Change Images: Photos are essential components of websites and you should start by modifying them the first time. The rule of thumb is that you should have an image that is perfect in dimensions and the image must always be compressed. If we have an image with 1200x1200 pixels. You want to use it for just 300x300 pixels. That is, the real space on the webpage is smaller than the actual dimensions of your image. Therefore, you must change the size of the image! Make use of this Online Compression tool to compress your image. You'll be amazed by the results A good tool to use is GZip from your host or you may want to search for an application.

Abolishing rendering-blocking JavaScript The way to format your website for when HTML begins loading means that the site requires loading CSS first, and then JavaScript after that.

CSS in content above-the-fold Same meaning as step 2.

Web Caching: Caching a page's content in memory allows you to quickly return to a webpage without downloading this page from the Web again. We'll utilize a powerful plugin named "Autoptimize" by Frank Goossens (futtta) to accomplish this. Check out the section below titled "Caching + HTML, CSS, & JS Optimization" for more details on this.

Cut your HTML codes: If your know how to handle it You can tidy up your HTML/CSS/JAVASCRIPT code. In the event that you don't, you should seek assistance to clear the HTML code.

Reduce Server Response Time It is necessary to call your hosting provider for this, and talk about the issue that you're getting. Sometimes, it is fixed immediately after you have optimized your website. Sometimes, it's your hosting service that should be in charge of this.

Authorize Compression: The majority experts SEO experts utilize GZip which is provided by their hosting provider => Make sure to turn on or enable compression. You may also utilize WordPress plugins, too. You can also access this website for no cost https://tinypng.com

Make it Mobile-Friendly: Self-explanatory. We'll go over this further.

This is an Google link to conduct an Mobile-Friendly Test on your website:

https://search.google.com/test/mobile-friendly

Images Optimization - Part I

Part 1 of optimizing and compressing the images you use on your site.

VERY IMPORTANT:

There are many beautiful and high-quality, free images on www.unsplash.com

I would also suggest that you go on www.similarweb.com as well type in your URL www.unsplash.com in order to locate similar services that are free.

Google tool: Google tool:

https://developers.google.com/speed/page speed/insights go to the Image Optimization Section and read the suggestions. It will show the URLs of your images with suggestions from the tool: mostly regarding resizing and compression. Important to be aware of.

In the GTMetrix tool: https://gtmetrix.com you will find the same suggestions in"Optimize images" in the "Optimize images" section.

There are three steps involved in images Optimization and Compression: First, take the original image and Resize it, then compress it. Upload it onto the servers.

Change the size of your photos: Discover what size is appropriate of your image. If an image appears tiny on your site however, it might be large. (If the suggested image size is 50X50, do not upload an image that is smaller, it can affect the Page Speed Optimization score as well as your overall SEO Score) You can open that image in a new tab and look at the actual size of the image. Example: If you open a brand new window that shows a Facebook icon that is located in the footer it displays 300x300 pixels, but the reality is that you need 60x60 . Therefore, it must be optimized. This is the first step you must do.

It is possible to use this tool www.photopea.com for alter the size of your photos. It's similar to an FREE online photoshop.

It's time to reduce your images.

Compress the images you have: should create three folders for your photos. They are called "Original Images", "Resized Images" and "Compressed Images". The process works similar to this take the originalResize it, then compress it and then Upload it on the web server.

Here's a tool you can make use of to compress photos: www.tinypng.com and If you're an developer, you may also utilize their API.

. When you are uploading images on your website frequently, you might want use a compression tool to your WordPress. If not, don't include any plugins to your website in the hopes of stopping your website from loading more slowly.

. Then, upload your pictures to your server.

Images Optimization - Part II

Once you're finished with the process of Image Optimization and Compression the next thing you have to do is test your images Optimization using similar tools and evaluate the old results against the latest ones:

www.gtmetrix.com

https://developers.google.com/speed/page speed/insights

Now you should have noticed that you should have noticed that your Page Performance Optimization Score is likely to increase and your pictures should be reduced in size. Simply look at the old image (before compressing and resizing) to the new image (after compressing and resizing) and observe the difference in speed. increased.

Let's get started by discussing the basics of what CDN is and how you can install a free CDN on your site.

What is CDN - Adding CDN for Free

In this course we'll discuss CDN. CDN is "Content Distribution Network" or "Content Delivery Network". Both are the same.

The content delivery network (CDN) can be described as a set of made up of servers, which are placed at various locations in order to accelerate the loading time of content, while providing it to a place close to the site's visitor. CDN can also assist us to create a secure website. CDN will be able to detect the traffic and tell you whether this is normal or high-volume traffic. CDN will also determine the authenticity of this traffic or fake traffic, for instance. The result is that CDN detects them all.

To learn more about CDN Let us look at two scenarios: one without CDN and the other one that has CDN:

Without CDN: Your server is located in The USA and you can see that there are three users online: 1 from India One from Japan and one who is from UK. Every request made by those users located in different countries will reach your server located in the U.S. This is a time-consuming process. That is it takes more time to receive a response. An extra millisecond WHICH MAY SEEM TO BE NOTHING BUT IN THE WORD OF CYBERSPACE, IT'S SOMETHING.

With CDN: When you've added a CDN it functions as a medium Between your Server and clients. In the sense that CDN is a CDN is a replica of your server which functions like a satellite on the local level to increase the speed of the process. It's an Cache of content Cache on your servers, which reduces the bandwidth. Thus, clients feel that the server is located in India for instance and not located in the USA. This is because the CDN is now locally installed The only thing that's local to India can be described as the CDN and not the server. It serves as a mediator between user and server, and is therefore a security medium.

Requests are directed via India instead of in the U.S. and this will decrease the time.

The question is: can small businesses offer CDN technology? The answer is no. A few large Guys provide free CDN+ www.cloudflare.com can be one. THIS IS GREAT. Particularly if you have a server located in the USA and your clients are spread around the world , for example.

Follow these steps to create an absolutely free CDN:

Create your account for free Cloudflare account as easy as possible add your site (include HTTPS if you have it) This could take a little longer. The next screen you'll be required to confirm your DNS records. Go to Next. offer 4 plans, and one of them is free. Choose Free Account. The goal is to change your current DNS records by using Cloudflare DNS data. => Copy Cloudflare DNS1 and DNS2.

Go to the hosting/domain Nameserver1 and 2 for your company, then copy Cloudflare DNS1 along with DNS2 on Nameserver1 and Nameserver2 and click "Update Name Servers".

It can be up to 24 hours for you to update your Nameserver. Visit Cloudflare and click "Recheck Nameservers" to see what has changed.

The next step is to continue to configure the CDN configuration/setup and then add a Free SSL Certificate.

If you're looking for an absolutely free SSL certificate for your site This is the perfect moment to obtain one. Follow these steps:

You want to check this interesting article: Get Free SSL For Your Website with Cloudflare - Easy Setup Guide: https://pushalert.co/blog/free-ssl-easy-setup-guide/

When your account is activated and you are logged in, you will see the green indicator that indicates that your website has been CDN enabled.

After your account has been active, your website will also receive a no-cost SSL Certificate if you enable it during the process of confiuration.

Check out this toolbar at the at the top of the Cloudflare dashboard: You will be able to look up your DNS, Analytics Speed, Caching speed and many more.

Concerning the speed Tab In the case of Speed Tab, if you're using an WordPress plugin to cache your pages, then you're all set. However, if you're not using WordPress or any other plugin, you can decrease the size of the web pages that contain source code in JavaScript, CSS, & HTML fields. Select the "Speed" tab (see picture above) and then go through it in the way you like. The aim is to reduce the size of all of the

HTML, JavaScript, and CSS files, and also to shrink the size of your files.

You might want to close the other tabs open as default.

Note:

* If you're working with the HTML platform, visit www.minifier.org and select the kind of file you'd want to reduce (HTML or js or CSS) Simply enter the URL address for that file. Click "Minify".

* If you are using WordPress then follow the instructions below to help you understand JavaScript, CSS, & HTML fields:

Visit your web site's WordPress Admin and install this plugin: "Cloudflare" by Cloudflare.

Once the application is activated, return the Cloudflare Dashboard and click Home> under "Optimize Cloudflare for WordPress" then click"Apply" "Apply" button.

It will also optimize your site's performance.

Now, go to Settings"Settings" > "Speed" and check ON "Always Online" => If your server is offline, Cloudflare will serve your static web pages from their cache.

Caching + HTML, CSS, & JS Optimization

After you are done with Cloudflare + SSL Installation, go to any SSL Checker such as https://www.sslshopper.com => Scroll down and then enter your website to check out its security.

Now, let's Install "Cache Plugin" on your WordPress website. The idea is that your site needs to be converted to Static files. We then have to reduce the size of those HTML, JavaScript, and CSS files to make them smaller. There are two plugins that you must be able to install, and activate:

1. "Comet Cache": This plugin creates a static image for your entire website. It continues to function automatically by

eliminating old files and generating new ones. It also creates files for your photos like converting larger '.png images to smaller '.webp types'. The plugin was developed by websharks.

2. "Autoptimize": This plugin optimizes your CSS and HTML. The plugin bundles them together into one file. The plugin not only reduces the files, but also reduces the number of requests". The plugin was created by Frank Goossens. The plugin can optimize other aspects too, such as fonts, fonts, etc.

After Comet Cache is installed =go to Settings and then click the "Yes, enable Comet Cache" radio button. You can set the other options as default. Comet Cache comes also with GZip Compression that you can activate to enable image compression. Click "Yes to enable GZip compression" under"Apache Optimization" section "Apache Optimization" section.

After "Autoptimize" is installed, the next thing you need you need to do is visit Settings > Autoptimize> > Here, you can have the option of optimizing the HTML code, JS, & CSS codes. Click on each of the buttons that corresponds to the above. Then click to "Save and Empty Cache" (this plugin can also be used to create an unchanging file and then empty your cache, too.).

Go to"Extra" and then click on the "Extra" Tab and select the options you prefer.

Select"Optimize More". Click on "Optimize More" Tab and examine other plugins that they include.

Be aware: The plugins we have installed, as well as other plugins can slow your website down for a time However, once they begin making improvements to the website it will be resolved. The speed may slow, but will rise later. I suggest running the test of speed on your site prior to and after adding an extension and comparing the scores before and after to determine if you would like to keep the plugin , or perhaps not. Make sure you uninstall a plugin the proper way. It can take time to adjust to the needs of your site. That is an application that slows your site now could speed up within a couple of days. Be patient here.

Now, we'd like to install a different plugin named: "Browser Cache Expiration" Search for your .htaccess file inside your cPanel. Click to modify the file. You will find a line on your website's structure like your WordPress as well as the GZip, Comet Cache, and so on. It is necessary to include this code in the .htaccess file. Include the final line of the image, which is about CSS header files. It's cut off in the picture:

This code will increase the speed of your website by caching. All images plus HTML + CSS files will remain to the browser for a month, based on the date you have included in your code (in this instance, it's 1 month. Check out the above code).

Sometimes, your normal .htaccess file might be replaced by the Cloudflare's .htaccess. This issue will be addressed in the future Read the following section which is called "Browser Cache Expiration" - Don't panic, there's a very SIMPLE solution to this.

After you're finished with all the cache settings, all optimization settings, and all minifications settings, you must test again/check:

Make use of to use the "GTMetrix" tool and Google "PageSpeed Insights" tools once time => compare the score you received with the old score. The PageSpeed Score is available in the near future! Your site should be loading quickly. Also, you should be aware you can see that JavaScript files are integrated into one file. This is the case applies to CSS files too. Also,"Requests" "Requests" should be decreased in the near future. The total size of your page also needs to be reduced and the page should take less time to load.

Tips: To determine whether your website is caching using the Comet plugin by logging out from WordPress and then open your site => check the source codescroll down to the bottom until you find"Comet Cache" "Comet Cache" comment in green.

After you have completed all these tests, it is necessary to start your website using "New Incognito" Window Mode to ensure that the browser is properly cache optimization.

If you are now in this situation you must recheck these two other items: "Browser Cache Expiration" and "Render Blocking Error".

Browser Cache Expiration This is the "Expiration date" of the cache. It should be listed in the .htaccess file. SEE THE PREVIOUS PICTURE ABOVE. If you see the error message on the SEO reports, that indicates the .htaccess file might be messing with the expiration date of your browser Cache. It is possible that Cloudflare has overwritten this file also. It's possible! To resolve this problem, log onto the Cloudflare dashboard and click"Caching" "Caching" Tab => Find"Browser Caching Expiration" "Browser Caching Expiration" date The default setting is it's set at 4 hours. You want to change it to a month, which is the same as what you have in the .htaccess file (see the previous image above). SIMPLE!

The definition of a render-blocking error is: The word "render" means loading, and, consequently, when something is rendering-blocking, it's because it's hindering the site from loading at the speed it ought to.

If you notice an error like this in the SEO report, it indicates that something is preventing the page from loading as it should. The reason for this is the case that your JavaScript is within your HTML header and not at the bottom. So you must add the JavaScript to the top.

TIP: You may want to add "Analyze Page Performance" Chrome Extension to measure site speed on the fly => https://chrome.google.com/webstore/detail/analyze-page-performance/hemibacgndhdhkfahkjdedjdgfapmfki?hl=en-US

To test it, simply go to the URL you wish to test. Hover the extension until you see the score. That's it.

Leverage Browser Caching - Page Speed Test After Optimization:

If you test your page's speed using GTMetrix It is likely you'll find that "Requests" are still high and your objective for this number should reduce it.

To resolve this issue you must log into your WordPress Admin and then go to the "Autoptimize" Plugin under Settings and then by the way, you'll want to record the changes in case you experience certain JavaScript or CSS errors , so you are able to reverse. Go to Autoptimize Settings:

> In the section "MAIN TAB": Check "ON" the following:

"Optimize HTML Code",

"Optimize JavaScript Code",

"Also aggregate inline JS",

"Add try-catch wrapping?" (pay close attention to this since you might encounter an error, and it could cause a breakage on your website. If you encounter an error, you must reverse it), "Optimize CSS Code", "Generate Data: URLs for images?", "Also "aggregate inline CSS?", "Inline All CSS?"